Arctic Sustainability Research

The Arctic is one of the world's regions most affected by cultural, socio-economic, environmental, and climatic changes. Over the last two decades, scholars, policymakers, extractive industries, governments, intergovernmental forums, and non-governmental organizations have turned their attention to the Arctic, its peoples, resources, and to the challenges and benefits of impending transformations. Arctic sustainability is an issue of increasing concern as well as the resilience and adaptation of Arctic societies to changing conditions.

This book offers key insights into the history, current state of knowledge and the future of sustainability, and sustainable development research in the Arctic. Written by an international, interdisciplinary team of experts, it presents a comprehensive progress report on Arctic sustainability research. It identifies key knowledge gaps and provides salient recommendations for prioritizing research in the next decade.

Arctic Sustainability Research will appeal to researchers, academics, and policymakers interested in sustainability science and the practices of sustainable development, as well as those working in polar studies, climate change, political geography, and the history of science.

Andrey N. Petrov is Director of the ARCTICenter and Associate Professor of Geography at the University of Northern Iowa, USA. IIis expertise is in economic geography and regional development in the Arctic, sustainability science, social indicators, impacts of development, benefit sharing, and sustainable economy in remote regions. He co-edited *Arctic Social Indicators: Implementation (2015)* and authored more than 40 publications on Arctic issues. Dr Petrov leads three NSF projects focused on resilience and sustainability. He also holds leadership roles in the International Arctic Social Sciences Association, International Arctic Science Committee and American Association of Geographers.

Shauna BurnSilver is an Environmental Anthropologist who studies how socio-economic and climate changes affect relationships between people and the natural resources they depend on. Key questions of interest are how households, communities, and stakeholders define concepts of "well-being", "progress", "development" and "sustainability" given the intersection of strong socio-economic and political changes and long-term cultural histories of land use in particular landscapes, and how these perspectives translate into decisions within mixed economies at the scale of households and communities that matter for sustainability outcomes. She has worked in Arctic Alaska, Inner Mongolia, China, and Africa.

F. Stuart Chapin III is Professor Emeritus of Ecology at the University of Alaska Fairbanks. His research addresses the effects of changes in climate and wildfire on Alaskan ecology and rural communities. He also collaborates with communities and agencies to develop options that increase sustainability of ecosystems and human communities over the long term despite rapid climatic and social changes. His research in earth stewardship explores ways that society can proactively shape changes toward a more sustainable future through actions that enhance ecosystem resilience and human well-being.

Gail Fondahl is a Professor of Geography at the University of Northern British Columbia, Canada. Her PhD is from the University of California-Berkeley. Professor Fondahl's research focuses on the legal geographies of indigenous territorial rights in the Russian North and the cultural and governance dimensions of Arctic sustainability. She serves as Canada's representative to, and Chair of, the International Arctic Science Committee's Social & Human Sciences Working Group. Dr Fondahl was President of the International Arctic Social Sciences Association (2011–2014) and remains on IASSA's governing council. She co-edited the second *Arctic Human Development Report* (AHDR-II).

Jessica K. Graybill is a Human Ecologist and Associate Professor of Geography and Director of the Russian & Eurasian Studies Program at Colgate University. Her focus in Eurasia is on the Far East and North where she conducts research related to resource use, extraction, and climate change in the Russian Far East and North. Currently, she is a co-principal investigator on two National Science Foundation grants in Arctic resilience and sustainability. She is also an interdisciplinary research methodologist. Recent publications include *Urban Climate Vulnerability and Governance in the Russian North* (2015) and *Cities of the World, Sixth Edition* (2016).

Kathrin Keil is a Scientific Project Leader at the Institute for Advanced Sustainability Studies (IASS) in Potsdam, Germany, where she leads the Arctic research project GloCAST (Global Change and Arctic Sustainable Transformations). GloCAST uses the Arctic as a prominent case to illustrate interrelations between global and regional change processes and between stakeholders from within and outside the Arctic. Kathrin received her PhD in Political Science from the Freie Universität Berlin. She is Senior Fellow of The Arctic Institute – Center for Circumpolar Security Studies and part of the German delegation to the Sustainable Development Working Group of the Arctic Council.

Annika E. Nilsson is Senior Research Fellow at Stockholm Environment Institute and Affiliated Faculty in Environmental Politics at KTH Royal Institute of Technology. Her work focuses on the politics of Arctic change, with research on environmental governance, communication at the science-policy interface, and the relationship between resource extraction and sustainable Arctic communities. She has participated in several scientific assessments under the auspices of the Arctic Council. Nilsson received her PhD in environmental science in 2007 following a 20-year career as a science journalist.

Rudolf Riedlsperger is a PhD candidate (Geography) at Memorial University of Newfoundland in St John's, Canada. His research interests revolve around sustainable community development. A particular focus is on sustainable public housing in Arctic and Subarctic regions, viewed both from a physical and sociocultural perspective.

Peter Schweitzer is Professor at the Department of Social and Cultural Anthropology at the University of Vienna and Professor Emeritus at the University of Alaska Fairbanks. His theoretical interests range from kinship and identity politics to human-environmental interactions, including the community effects of global climate change, and his regional focus areas include the circumpolar North and the former Soviet Union. Schweitzer is past President of the International Arctic Social Sciences Association, past Chair of the Social and Human Sciences Working Group of the International Arctic Science Committee, and currently the Director of the Austrian Polar Research Institute.

Routledge Research in Polar Regions
Series Edited by Timothy Heleniak

The Routledge series in *Polar Regions* seeks to include research and policy debates about trends and events taking place in two important world regions: the Arctic and Antarctic. Previously neglected periphery regions, with climate change, resource development and shifting geopolitics, these regions are becoming increasingly crucial to happenings outside these regions. At the same time, the economies, societies and natural environments of the Arctic are undergoing rapid change. This series seeks to draw upon fieldwork, satellite observations, archival studies and other research methods which inform about crucial developments in the Polar regions. It is interdisciplinary, drawing on the work from the social sciences and humanities, bringing together cutting-edge research in the Polar regions with the policy implications.

New Mobilities and Social Changes in Russia's Arctic Regions
Edited by Marlene Laruelle

Climate, Society and Subsurface Politics in Greenland
Under the Great Ice
Mark Nuttall

Greenland and the International Politics of a Changing Arctic
Postcolonial Paradiplomacy between High and Low Politics
Edited by Jon Rahbek-Clemmensen and Kristian Søby Kristensen

Arctic Sustainability Research
Past, Present and Future
Andrey N. Petrov, Shauna BurnSilver, F. Stuart Chapin III, Gail Fondahl, Jessica K. Graybill, Kathrin Keil, Annika E. Nilsson, Rudolf Riedlsperger, and Peter Schweitzer

Arctic Sustainability Research
Past, Present and Future

Andrey N. Petrov, Shauna BurnSilver,
F. Stuart Chapin III, Gail Fondahl,
Jessica K. Graybill, Kathrin
Keil, Annika E. Nilsson, Rudolf
Riedlsperger, and Peter Schweitzer

Routledge
Taylor & Francis Group
LONDON AND NEW YORK

First published 2017 by Routledge

2 Park Square, Milton Park, Abingdon, Oxfordshire OX14 4RN

52 Vanderbilt Avenue, New York, NY 10017

Routledge is an imprint of the Taylor & Francis Group, an informa business

First issued in paperback 2019

British Library Cataloguing in Publication Data
A catalogue record for this book is available from the British Library

Library of Congress Cataloging in Publication Data
A catalog record for this book has been requested

ISBN: 978-1-138-08830-6 (hbk)
ISBN: 978-0-367-21910-9 (pbk)

Typeset in Times New Roman
by Swales & Willis Ltd, Exeter, Devon, UK

Contents

Figures

Acknowledgments

This book was co-sponsored by the International Arctic Science Committee, International Arctic Social Sciences Association and NSF RCN Arctic-FROST (PLR # 1338850) and ASUS (PLR # 1532655). Special acknowledgment to Ann Crawford and Susan File for administrative support. We are also indebted to reviewers and participants of discussion forums at ASSW-2015, IGU-2015, and Arctic-FROST annual meetings (2015 and 2016).

Note: this book does not represent the views of the National Science Foundation.

IASC

IASSA
INTERNATIONAL ARCTIC SOCIAL
SCIENCES ASSOCIATION

1 Background and purpose

The Arctic is one of the world's regions most threatened by ongoing and increasing cultural, socio-economic, environmental, and climate changes. Over the last two decades, multiple stakeholders – scholars, policymakers, extractive industries, local-global governments, local and indigenous communities –have turned their attention to the Arctic, its peoples and resources, and to challenges and benefits of impending changes. The International Conference on Arctic Research Planning (ICARP) has now convened three times, most recently in April 2015. Of increasing concern within the Arctic and within ICARP endeavors is addressing sustainability. This book based on the ICARP III white paper, aims to give a comprehensive understanding of the current state of research on sustainability and sustainable development in the Arctic and identifies related knowledge gaps and research priorities for the next decade, until ICARP IV. Here, we provide a historical overview of sustainability concepts in global and Arctic contexts, a progress report on Arctic sustainability research, and recommendations for prioritizing research. To evaluate the state of sustainability research in the Arctic, we reviewed science plans from ICARP II (2005) and progress since then in addressing sustainability and sustainable development research. Particularly, we analyzed progress towards Science Plans 1, 2, 10 and 11, focusing on economic development, indigenous considerations, social–ecological change, and research processes and communication, respectively. Here, we identify future research priorities through this review of research on sustainability and sustainable development in areas of interest identified by ICARP II and areas that have emerged since 2005. To narrow and deepen our analysis, we especially focus on theory, methodology, synthesis, indicators, governance, and ecological dimensions. Written by a team of anthropologists, ecologists, geographers, and political scientists, our analysis is interdisciplinary in scope. The first draft, prepared by eight researchers, resulted from an interactive four-day workshop organized around sessions broadly representing the different sections of this white paper. This draft was then shared with a broader network of researchers, either as contributors or reviewers, and presented to the academic public for discussion and feedback at several venues[1], including the ICARP III meeting in Toyama, Japan in April 2015. The summary of the key findings was published in 2016 (Petrov 2016) and we have since received additional comments that we incorporated in this book.

Note

1 Arctic Science Summit Week/International Conference on Arctic Research Planning III, Toyama, Japan, April 2015.

2 A brief history of sustainability as a concept in the Arctic and beyond

One might argue that sustainability thinking emerged almost as early as humans started to roam the earth, as they relied on the use of natural resources while noticing the negative effects of overuse in some instances. Still, the discourse of sustainability and sustainable development is largely a 20th century phenomenon, with antecedents reaching back a few centuries. While the aim of this book is not to provide a comprehensive overview of the history of sustainability, we offer a brief overview of the development of conceptualizing sustainability. We will touch upon pre-20th century conceptualizations and discuss some of the global 20th century developments before focusing on the Arctic since the 1980s.

2.1 Conceptual beginnings in "Western" thought and early nature protection

The term sustainability may be traced back as far as the 1650s and the so-called Hausväterliteratur or "head of household literature," which originated in best practice books where "the bottom line . . . was the maintenance of household income" (Warde 2011: 153–154). An early sustainability discourse can also be found in the literature about forestry from the 17th and 18th centuries. For example, the work Sylvicultura oeconomica, a forestry manual published in 1713 by Hanns Carl von Carlowitz, used the word "nachhalthende", which can be translated as "sustainable." While this focus on forestry might seem difficult to comprehend today, maintaining forest productivity and the supply of wood was crucial for energy supply before the era of fossil fuel (Caradonna 2014: 32).

Thus, one could argue that at the beginnings of the "sustainability" discourse stands the necessary integration of ecological and economic dimensions – the forest, the farm. The household can only survive and blossom if income and resource availability remain in balance. The third "e" of today's sustainability discourse (in addition to "e"conomy and "e"cology) – namely equity – appears much later. The three "es" are often described as environmental, economic and social "pillars" of sustainability (United Nations 2002; Kates et al. 2005). In today's discourse, sustainability integrates social dimensions, including aspects related to human development but also to equity. Equity in this context implies a fair distribution of benefits and a discourse that has its roots in tensions related to colonialism and decolonization. An additional forerunner to the discussion on the relationship between equity and sustainability was the "too-many-people" narrative, most famously expressed by Thomas Malthus' *An Essay on the Principle of Population* (1798) and leading to 20th century contributions such Paul Ehrlich's *The Population Bomb* (1968).

Another thread in the roots of today's sustainability thinking comes from the preservation movement, which can be traced back to the second half of the 19th century and found its most iconic expression in the creation of U.S. National Parks under President Theodore Roosevelt. The philosophical underpinning for these developments was provided by people like Henry David Thoreau (1817–1862); who brought novel perspectives on human-environmental relations, and in his most famous book, *Walden, or Life in the Woods* (1854), which advocated a simple life in harmony with nature.

In Russia some of the ideas that shaped the view of sustainability came from the works of Vladimir Vernadsky (1945), Nikita Moiseev (2000) and their followers who advocated the "rational," science-driven development that will inevitably lead to a sustainable future. The concept of the "noosphere" (a "mind-sphere" governed by human ingenuity) advanced by Vernadsky was considered by him an evolutionary stage resulting from transforming the Earth's biosphere (a sphere of life) by the humans.[1]

International sustainable development politics have one origin in the international nature preservation movement. This movement started to gain momentum in the early 1900s, and focused on the protection of iconic species in colonial territories (Linnér 2003:30), including attention on the Arctic with protection of musk ox in Greenland and reindeer in Spitzbergen (Conwentz 1914). Toward the end of World War II, a parallel global discourse began to emerge that was more closely linked to the notion of long-term productivity, especially in agriculture (Linnér 2003). Food and agriculture had been addressed by international society in the League of Nations (forerunner of the United Nations) since the 1920s, but a conference in 1943 led to the creation of the Food and Agriculture Organization (FAO). Its goal was to recommend national and international action toward "conservation of natural resources" and "improved method for agricultural production" (Mayne 1947). This focus on natural resources was closely tied to growing political concerns about producing enough food and other resources for the growing world population. It has been described as the emergence of a new world order and a new way of viewing the relationship between humans and nature (Linnér 2003: Ch. 2). A key event in the institutional development of this discourse was the United Nations Scientific Conference on the Conservation and Utilization of Natural Resources in 1949, which focused on food production, population growth and the use of science to ensure effective use of the resources to raise the standard of living (Aull et al. 1950; McCormick 1991).

In the mid to late 1940s, one can thus see a merger of two parallel discourses that have later played a central role for the concept of sustainable development: conservation of nature and the use of nature-based production systems to support basic human needs. Both perspectives were becoming formalized in organizational structures and institutional norms that were increasingly addressing politically relevant issues related to development, poverty and human well-being. The organizational developments were also closely linked to the emerging international governance system associated with the United Nations, which provided an arena for government voices with different priorities to the conservation movement.

At this time, concern about pollution and its potential impacts was not yet much on the political agenda. This started to change in the mid-1950s with protest against nuclear weapons tests. In the 1960s, with the publication of Rachel Carson's *Silent Spring* (1962), attention turned to chemical pollution. Together, these two developments laid the foundation for the environmental movement in the West (Worster 1985), and to a lesser extent in the Soviet bloc (Pryde 1991; Feshbach and Friendly 1992; Graybill 2007). The issues in focus were highly relevant to the Arctic, where the military used many of the chemicals that are now known to be both toxic and persistent in the environment, such as DDT to deal with mosquitoes and other insects, and PCBs as integrated in the military installations that were central during the Cold War, including the DEW line stations (AMAP 1997).

The emerging linkage between environment and development also affected discussion within the international conservation movement, especially within IUCN (International Union for the Conservation of Nature). The idea that protected areas and threatened species were best safeguarded if local people also felt it was in their best interests received greater attention (Christoffersen 1997; Selin and Linnér 2005). In 1980, the approach was formalized in the World Conservation Strategy (IUCN 1980), which included explicit discussion of "sustainable development" as a concept. The strategy was the result of collaboration with a wide range of international organizations and the process initially was part of a program called Conservation for Development, which was intended to link the organization's own scientific expertise in conservation with the work of development agencies (Louafi 2007). As described by Selin and Linnér (2005), the IUCN's first version of the strategy, which focused mainly on conservation, met strong opposition from African members of organizations who wanted to include both conservation and development, which was also a priority of the United Nations Environment Programme (UNEP). Later work on the strategy involved a wider range of organizations and came to stress the mutual dependence of conservation and development. The World Conservation Strategy introduced the concept of sustainable development to a broader international public and became an important inspiration for the World Commission on Environment and Development, called the Brundtland Commission (WCED 1987; Christoffersen 1997; Selin and Linnér 2005).

The World Conservation Strategy also influenced the indigenous movement, and the Inuit Circumpolar Council (ICC) officially endorsed the document (Keith and Simon 1987). In this context, industrial development was also part of the discussions. While welcoming the economic development that would come with industrial activities, Keith and Simon emphasized that large scale, quickly constructed industrial projects did not fit well with how the Inuit saw sustainability. Rather, they wanted to see an emphasis on duration of benefits, distribution of benefits and costs, and influencing the decision making process. Relating to issues of equity and fate control, Keith and Simon (1987) used Saami protests in Alta, Norway, against a major hydroelectric project as an example of how this lack of influence spurred conflict. In addition to introducing the concept of sustainable development, the World Conservation Strategy featured several other concepts of relevance for contemporary discussions. One of its main aims was "to maintain essential ecological processes and life support systems," echoing development within the science of ecology towards ecosystem-based thinking. Visible in the text is both an emphasis on moral obligation towards future generations and practical needs of "securing supplies of food, fibre and drugs" (IUCN 1980: Ch. 3 point 11). The World Conservation Strategy was a conceptual document without legal status. Its major implementation was conceived to take place at the national level. Nevertheless, Louafi (2007) has argued that its diverse membership of scientists, non-governmental organizations, and governments created an epistemic community that became instrumental in defining a new foundation for international governance related to the environment. The Strategy can be described as an exercise that brought together actors with different interests and worldviews on the role of conservation in a context that could handle emerging conflicts by emphasizing a new, broader framing. In this sense, sustainable development can be seen as a boundary concept with the same role as a boundary organization (Guston 2001) for bringing diverse views together and creating a context for dialogue.

When in March of 1987 the World Commission on Environment and Development, led by Gro Harlem Brundtland (then Prime Minister of Norway), published a report titled "Our Common Future," the concept of sustainable development became popular among a wider audience. The report conceptualized sustainable development as "development that meets the needs of present generations without compromising the ability of future generations to meet their own needs" (WCED 1987). The report provided a framework that brought together the natural or environmental limits of development with the potential for new directions in social development contained within those natural limits (Chance and Andreeva 1995).

Spurred by Brundtland, at the global level the focus on sustainable development became institutionalized with the UN Conference on Environment and Development (UNCED) in Rio de Janeiro in 1992 with its Agenda 21. The conference also brought a new international governance landscape, with two conventions of particular importance to the Arctic: the Climate Convention and the Convention on Biological Diversity.

2.2 Sustainability in the Arctic

To an extent, the history of conceptualizing sustainability and sustainable development in the Arctic parallels the processes described thus far and closely relates to both environmental protection and stewardship, and economic development. In addition, socio-cultural well-being of indigenous communities is a key consideration when the term of sustainability is used in the Arctic context, particularly in relation to work within the Arctic Council (Tennberg 1998; Keskitalo 2004: National Academies 2014). In the Arctic, sustainability and sustainable development are inextricably linked to resource exploitation. Compared to other parts of the world, colonial resource extraction is a relatively recent phenomenon in many parts of the Arctic (e.g., the late 19th/early 20th century in Arctic North America) (Hacquebord 2012; Agranat 1992; Armstrong 1978; Rea 1968), even though economic development is a recurring practice and commodities such as whaling, fur and gold historically initiated resource boom and bust cycles. While early industrial development as part of building modern nations transformed both the physical and social landscapes in parts of the Arctic, e.g. mining in Northern Sweden (Sörlin 1988) and the industrialization of the Soviet North (Agranat 1992) it did not lead to major protests. However, when megaprojects in northern development started to transform "hinterlands" into large scale industrial nodes in the 1950s (Bone 2009), responses were set in motion, both inside and outside of the Arctic, to counteract environmental degradation and the disenfranchisement of indigenous communities. These responses were in large part due to lax or non-existent regulation regimes as well as the remoteness of development sites (McCannon 2012).

Throughout the Arctic, but especially in North America and Fennoscandia, indigenous activists protested megaprojects and their negative sociocultural and environmental impacts. Moreover, opposition against nuclear weapons tests served as an early rallying cause. An example where the two issues combined was the battle against "Project Chariot" in Alaska, where the proposal to create an artificial harbor through the "peaceful" use of atomic detonations contributed to the self-organization of indigenous political interest (O'Neill 1994; Daley and James 2004). Another, more comprehensive, example of local or indigenous organization against economic development is seen in sustained reaction to the expansion of energy production in the North, including both the building of hydroelectric dams and a wave of interest in Arctic hydrocarbons following on the OPEC (Organization of the Petroleum Exporting Countries) oil embargos and oil price crises of the 1970s. Together these developments became important elements in the formation of indigenous political movements in both North America and Fennoscandia. For example, in Norway, plans for exploiting the Alta River led to intensive protests organized by Folkeaksjonen mot utbygging av Alta-Kautokeinovassdraget, including civil disobedience actions starting in 1979 (Paine 1982; Andersen and Midttun 1985). These protests merged environmentalism with Saami rights activism, and Alta was in many ways the start of the Saami political movement in Norway. In Canada, hydropower and its infringement on native land was a key issue in the negotiations regarding a dam in James Bay, Quebec, which led to the first land claims agreement since the 1920s, involving Quebec Cree Indians and Inuit and Naskapi Indians (Richardson 1977; Feit et al. 1995; Peters 1999; Tanner 1999). In the North American Arctic the focus was also on oil, especially Prudhoe Bay, which stirred the debate about Alaska's land and resources and eventually led to the Alaska Native Claims Settlement Act (Berger 1985; Mitchell 2001; Shadian 2014; Stuhl, 2016). In Canada, attention turned to plans for building a large pipeline through the Mackenzie Delta, which led to the Berger Inquiry between 1974 and 1977 with hearings in 35 different communities in the Mackenzie Valley and all major cities in Canada (Gamble 1978; Berger 1977; Sabin 1995; Shadian 2014). The inquiry highlighted environmental, economic, and social impacts and recommended that the pipeline should not be built. In a text from 1987 Keith and Simon describe how these various plans for industrial development "pushed the hitherto unorganized northern peoples to organize and protest" (Keith and Simon 1987, 211). In relation to the Inuit, the details of this dynamic have been described in detail by Shadian in her book *The Politics of Arctic Sovereignty* (2014).

Land claims negotiations and self-government agreements ensued in North America, in part as a response of indigenous populations against the negative environmental and sociocultural impacts outlined above. In Fennoscandia, the development of indigenous land rights has also been much slower than in North America, with the Finnmark Act from 2005 as the only major legal agreement that specifically provides for indigenous (Saami) influence on the use of land and resources (Poeltzer and Wilson 2014). It was a result of a 25-year long process that started with the Alta protests. In the Russian Arctic land claims agreements as such do not exist, and legislation protecting the rights of indigenous northerners to their homeland is less advanced and effective (Fondahl 1997; Fondahl and Poelzer 2003; Kryazkhov 2010, 2013). However, there have been other avenues through which northern indigenous groups have been able to at least partially address issues related to development and environmental protection: for example, the ICC was founded in 1977 with the mission to "preserve the Arctic environment" (Shadian 2010, 487). In the Arctic there are varying negotiation mechanisms from formal impact and benefit agreements to informal agreements between companies, state and communities (Novikova 2010; Tysiachniouk 2010; Sosa and Keenan 2001). In the Russian Arctic environmental and ethnological assessment processes have provided some recourse for local concerns, while also facilitating industrial access (Novikova 2008). While efforts to gain indigenous rights and further land claims have had some successes, Usher (2003; and see Nuttall 2010) have emphasized that these processes of engagement may also be seen as a means for state governments to remove barriers to development projects. In other words, land claims agreements also served (and continue to serve) as a means of assimilating indigenous communities into capitalist societies through socio-economic development (Saku 2002; Hitch and Fidler, 2007).

In October of 1987, months after "Our Common Future" was published (see Section 2.2), then leader of the USSR Mikhail Gorbachev suggested a series of policy initiatives, a step identified as the beginning of the end of the Cold War in the Arctic (Åtland 2008). As Young (2010, 168) states, Gorbachev's iconic Arctic "zone of peace" speech had the effect "to decouple or delink the Arctic from overarching global concerns and to bring down barriers that had precluded efforts to create co-operative arrangements encompassing the Arctic as a distinct region." Gorbachev's speech helped to emancipate the Arctic from security interests through proposing initiatives revolving around military, economic, societal, and environmental areas, and including indigenous peoples (with an emphasis on cultural connections), scientific cooperation (which made possible the foundation of IASC in Stockholm in 1990), environmental cooperation, and energy cooperation (focusing primarily on more efficient and environmentally sound extraction of resources), among others (Åtland 2008).

Brundtland and Gorbachev initiated two separate processes. The first process introduced the concept of sustainable development to a broad, global audience. It also emphasized the important role of indigenous groups in the context of sustainable development (WCED 1987), an aspect that is of particular importance when we look at sustainable development in an Arctic context. The aim of the second process was to foster Arctic collaboration and cooperation through suggestions that were compatible with the sustainable development considerations put forward by Brundtland. In his Murmansk speech Gorbachev even urged the public to "applaud the activities of the authoritative World Commission on Environment and Development." While his speech had important geopolitical consequences, shifting the paradigm of state interaction "from confrontation to cooperation" (Heininen 2011), Gorbachev's call for collaboration also paved the way for implementing policies fostering sustainable development. As we will outline below, the Arctic Environmental Protection Strategy (AEPS) brought these two processes together.

Sustainable development had become an explicit part of the Arctic research agenda in the late 1980s, for example in the Northern Sciences Network of UNESCO's Man and Biosphere Programme (MAB) (Archer and Scrivener 2000). Its integration into international environmental politics in the early 1990s naturally also affected the rapid development of circumpolar Arctic politics in connection with the negotiation about circumpolar cooperation in the so-called Rovaniemi process. The 1991 Rovaniemi Declaration that established the Arctic Environmental Protection Strategy (AEPS) mentions sustainable economic development and that the onus and rationale for the cooperation would be on avoiding unacceptable ecological and cultural impacts and on sustainable utilization of natural resource (Declaration 1991). However, in the end the focus was placed mainly on pollution control and human health (Tennberg 1998). Young (1998) comments that, despite the preambular language about sustainable development, global linkages and Arctic ecosystems, the document left out global issues such as climate change and ozone depletion. He also points out that the focus on environmental protection made it difficult for the AEPS to address the broader impacts of industrial activities in the Arctic.

The Ottawa Declaration, by which the Arctic Council was founded in 1996, includes affirmation of a commitment to "sustainable development in the Arctic Region, including social and economic development, improved health conditions and cultural well-being" (Arctic Council 1996). According to Keskitalo (2004), the sustainable development terminology that had initially been part of the negotiations for the AEPS was reintroduced mainly by Canada and the Inuit Circumpolar Conference (ICC)[2]. This was done to emphasize the utilization of resources, in contrast to conservation as the major goal, and was also connected to skepticism against the use of environmental arguments to prevent traditional hunting practices in the Arctic (Keskitalo 2004). In 1998, the Arctic Council established a Sustainable Development Program with the goal to "propose and adopt steps to be taken by the Arctic States to advance sustainable development in the Arctic, including opportunities to protect and enhance the environment, and the economies, cultures and health of indigenous communities and of other inhabitants of the Arctic, as well as to improve the environmental, economic and social conditions of Arctic communities as a whole" (Arctic Council 1998). In practice, the Sustainable Development Working Group (SDWG) of the Arctic Council has focused on a range of projects aimed at analyzing social aspects of human development in the region. Often, the SDWG has come to represent the "human dimension"[3] of issues treated in the other working groups.

The shift from environmental to broader sustainability concerns in the Arctic cooperation was closely connected to the increasing involvement of indigenous peoples' voices in Arctic politics. It thus mirrored a similar shift in discourse in international conservation politics that took place in the 1980s, from a species protection and conservation focus to broader concerns about human-environment relationships and on finding approaches that could gather greater political support both locally and internationally (Christoffersen 1997). But we can also see specific Arctic features of the discourse, such as concern that economic development would only be connected to industrial activities while actors outside the region would interfere with subsistence-based livelihoods and the possibility of basing economic development on products from these practices (Keith and Simon 1987; Keskitalo 2004).

At the working group level of the Arctic Council, different aspects of sustainable development continued to be compartmentalized until climate change started to enter the agenda with the Arctic Climate Impact Assessment (ACIA 2004; ACIA 2005). Based on this seminal report, impacts of climate change were no longer a potential future concern, but an issue that had to be addressed in the here and now. Moreover, the report gave a human face to the issues, especially the presentation of indigenous observations of climate change, in contrast to an earlier focus on abstract climate models (Nilsson 2009a). It was also starting to become clear that impacts of climate change could not be understood without an appreciation of the complex relationships between ecological and social contexts in different areas of the Arctic. There were some initial efforts to propose more integrative approaches. This is most visible in ACIA's chapter 17: Climate Change in the Context of Multiple Stressors and Resilience (McCarthy et al. 2005). The need for integrative approaches has been further reinforced with increasing political attention to the need for adaptation, where the Arctic Resilience Report (Arctic Council 2016) and the project Adaptation Action for a Changing Arctic are cases in point.

2.3 Indigenous/local concepts of sustainability and sustainable development

Some scholars of Arctic matters criticized the Brundtland Commission for viewing sustainable development as a simple compromise between environmentalism and economic growth, while neglecting important social dimensions of sustainability (Chance and Andreeva 1995) and the important tradeoffs that may be implied in pursuit of these goals. Crate (2006) noted that the Brundtland report confirmed "a dominant, western top-down economic worldview that bases ecosystem management on generalized prescription rather than specific contexts" (Crate 2006, 295). Instead of prescribing generic solutions, it is indeed important to consider what the processes of sustainability and sustainable development are, by whom they are carried out, and through which means they are being accomplished (e.g. Howitt 2001). When the concept of sustainable development was introduced, Keith and Simon (1987) asserted early on that development, "sustainable" or not, implied certain ideologies or processes that were fundamentally problematic, including the large-scale scope and rushing pace to realize economic projects such as pipelines, mines, and dams. Indeed, how key terms such as sustainability are translated into Indigenous languages can very much color how Indigenous people understand and respond (e.g. Graybill 2009; Cameron et al. 2015). The same issue is also relevant for related concepts, such as "climate change," "resilience" and "adaptation." For Inuit, concepts of sustainability and sustainable development are closely linked to questions of sovereignty. Indigenous knowledge (and ontologies?) speak to sustainable or considerate resource use (McGregor 2004), but in Arctic regions such knowledge is increasingly lost or impeded (Pearce et al. 2011). The Inuit Circumpolar Council (ICC) published a "Circumpolar Inuit Declaration on Sovereignty in the Arctic" as a reminder of the importance that "the rights, roles and responsibilities of Inuit are fully recognized and accommodated" in matters including climate change and resource development (ICC 2009). In particular, the declaration states: "In the pursuit of economic opportunities in a warming Arctic, states must act so as to:

(1) put economic activity on a sustainable footing; (2) avoid harmful resource exploitation; (3) achieve standards of living for Inuit that meet national and international norms and minimums; and (4) deflect sudden and far-reaching demographic shifts that would overwhelm and marginalize indigenous peoples where we are rooted and have endured" (ICC 2009)

There have been important local and regional processes that allow incorporation of specifics of particular places and spaces. Inuit Quaujimajatuqangit (IQ) offers one such example. IQ reflects alternative ways of thinking about how we interact with the environment, also in the light of global environmental problems (Wenzel 2004). It pertains to all aspects of Inuit culture, including values, worldview, language, social organization, knowledge, life skills, perceptions, and expectations (Tester and Irniq 2008). The concept emerged within the inception of the Government of Nunavut in 1999 as an attempt to move beyond the somewhat limited concept of traditional ecological knowledge. IQ does not adopt western language related to sustainability and sustainable development, but develops its own particular terminology. Important features are the incorporation of values (what is important to individuals and communities?) and cosmologies (what is the origin of our physical and social environments?). The principles and meanings of IQ are largely consistent with sustainable notions of human-environment relations, in part because they strongly emphasize the importance of social, cultural, environmental, and economic dimensions of sustainability (Ibid.).

2.4 Towards Arctic-based discourses of sustainability

The 2007 sea ice minimum made it even more apparent that climate change is a different type of environmental challenge than those that the Arctic Council had dealt with in the past (Christensen et al. 2013). The receding sea ice was not only discussed as an environmental challenge, it also became a starting point for a renewed attention to the potential for resource extraction and for potential conflicting cultural, social, economic, and political interests. In many ways the Arctic discourse has moved from cooperation and low politics to competition and high politics (see Nilsson 2012). Geopolitics and security concerns started to be discussed more openly and by a large range of actors both from within and from outside the region (e.g. Baev 2007; Berkman and Young 2011, Kraska 2011; Chaturvedi 2012; Heininen 2013; Tamnes and Offerdal 2014). Sustainable development is still part of the discourse, but other ways of framing the challenges ahead are also coming to the fore. They include a renewed interest in different ways of looking at security (Hoogensen Gjørv et al. 2014) as well as attention to vulnerability and resilience, which are discussed further in Chapter 4. Understanding how these concerns and discourses relate to each other is a major social science research task for the coming decade.

Overall, it has become clear that Arctic discourses on sustainability and related matters cannot be understood without their global contexts and the specifics of the Arctic region – such as politically engaged indigenous groups, fragile ecosystems undergoing rapid change, world-wide attention to large tracts of land and sea, etc. – now inform the global discussions. Thus, understanding Arctic sustainability becomes a prerequisite for understanding sustainability and sustainable development on Planet Earth.

Notes

1 Cf. *The Artic in the Anthropocene* (National Academy . . . , 2014).
2 Later renamed Inuit Circumpolar Council (ICC).
3 Some Arctic social scientists oppose the term "human dimension" (Stammler 2010).

3   ICARP II Science Plans

Reflection and assessment

A goal of this book is to consider what has been achieved in research on Arctic sustainability to date and to identify and identify key priorities for research for the coming decade, as part of the ICARP III process. The Second International Conference on Arctic Research Planning (ICARP II) was held in Copenhagen, Denmark in November of 2005. The event sponsored by IASC gathered more than 450 scientists, policy makers, research managers, representatives of indigenous peoples, and other stakeholders in Arctic research. One of the primary meeting outputs were twelve Science Plans focused on long-term research planning. Following the conference, plans were further refined to reflect input from the conference and online discussions.

In our review, we first turned to the ICARP II reports, which we examined for priorities related to sustainability research. We then considered to what extent recommendations for such research have been fulfilled. ICARP II did not articulate an integrated comprehensive plan for research in Arctic sustainability. Rather it included several documents that spoke of elements of sustainability and sustainable development from social and economic perspectives, in addition to several documents addressing biophysical sciences. For the purposes of this book we selected four ICARP II Science Plans that most closely correspond to sustainability science or research, as defined for the purposes of this book: Science Plans 1, 2, 10, and 111. Below we provide a brief reflection on each of the plans by comparing key proposed priorities with outcomes. Our survey is not exhaustive: our purpose rather is to assess the overall level of progress on these priorities and evaluate their relevance in respect to current and future directions of Arctic sustainability research.

3.1 ICARP II Science Plan 1. Arctic economies and sustainable development

Science Plan 1 (SP1) was the only one with explicit focus on sustainable development in the title and touched upon by many of its elements, emphasizing economic and social aspects of sustainability. The focal questions of SP1 were "How do Arctic economies work and how are they linked to issues related to sustainable development in general and to human development of Arctic residents and communities in particular?" Its premise was that "sustainable development requires the ability to set and achieve multiple social and environmental objectives in the context of change, through the management of human behaviour" (Huntingdon et al. 2005, 1). Using these questions as a starting point, the document proceeded to identify specific actions and research activities that would serve this goal in the next ten years. As a result, SP1 contained a large number of proposed approaches (or recommendations) addressing specific research questions.

The first key scientific question identified by SP1 relates to the meaning of sustainable development. The plan calls for a compilation and analysis of the terms equivalent to "sustainable development" in various public documents, policy papers, and Arctic languages and locations. In the last ten years some work has been done on this topic and a number of case studies have given insights into local understandings (e.g., Crate, 2006, 2008; Kruse et al., 2004; Riedlsperger et al. 2017), but no comprehensive study of the evolution of the concept as applied in the Arctic has been performed. Another important issue is the translation of "sustainable development" into various languages. In Russian, for example, "sustainable" could be translated as "immobile" or "stable," thus changing the original meaning of the English term (Danilov-Danilyan 2003).

Another SP1 direction was to perform an assessment of the goals of sustainable development in various contexts and comparison of perspectives and roles of various actors. To some extent these objectives were met by a number of successful comparative projects, even if sustainability has not been the explicit focus. An example is the Community Adaptation and Vulnerability in the Arctic (CAVIAR) project, but challenges arose with achieving comparability (Hovelsrud and Smit 2010). Considerable progress was achieved in investigating the understandings of sustainable development by indigenous groups, including Arctic Council Permanent Participants (Hovelsrud and Smit 2010). On the other hand, substantial work remains to be done to study perspectives of other key actors, especially industry and municipalities, as well as non-indigenous northerners. The role of cultural continuity, socio-cultural context and identity is also an important consideration in conceptualizing sustainability that needs further study. These efforts are visible in the forthcoming reports from the Arctic Council project Adaptation Actions for a Changing Arctic (AACA).

SP1 also called for special attention to investigating forces of sustainable development. In this respect, significant progress was made in identifying factors affecting sustainable development from vulnerability and resilience perspectives (see Chapter 4 for additional discussion); more recent literature focuses on exploring factors that affect adaptation (Hovelsrud and Smith 2010) and adaptive and transformative capacities (Kofinas et al. 2013), and interpreting these influences within the theoretical framework of social–ecological systems (SES) (Chapin III et al. 2009). Another focus has been to identify drivers of change that are relevant for the Arctic, including social and economic development (Cornell et al. 2013; Andrew 2014).

Although SP1 advocated developing a "rigorous definition" of sustainable development, scholarship in the last decade moved away from focusing on developing a single comprehensive ("best") definition in favor of reconceptualizing sustainable development as a process rather than outcome, thereby allowing a pluralization of understandings by focusing on local narratives of sustainable development (see Chapter 5).

Another key theme emphasized in SP1 relates to indicators for sustainable development. The Plan suggests effectively using available data, identifying measures relevant for assessing sustainable development and performing analysis of impacts of development using these indicators. Since 2005 some progress has been achieved in developing indicators and assessment of some sustainable development elements. Main accomplishments include the Arctic Social Indicators project (Larsen et al. 2010), the second Arctic Human Development Report (Larsen and Fondahl 2014), the Arctic Observing Network Social Indicators Project (AON-SIP) (Kruse et al. 2011) and the Arctic Marine Shipping Assessment (Artic Council 2009). Case studies that used indicators have also been conducted (e.g. Larsen et al. 2015; Vlasova and Volkov 2013). However, no comprehensive, integrated indicators frameworks that consider social and environmental variables have yet been developed for the Arctic as a whole (Ozkan and Schott 2013).

Modeling of sustainable development in Arctic communities has also achieved some progress. This is especially the case in scenario-based modeling (quantitative and qualitative), where several projects developed modeling frameworks to evaluate outcome of changes in Arctic communities (see Berman et al. 2004, 2017, Kruse et al. 2004, Lovecraft and Eicken 2011 for Alaska and Canada examples). Much has been done to study resource-based economies in various Arctic regions (Graybill 2013a; Graybill 2013b; Eilmsteiner-Saxinger 2011; Huskey 2006; Keeling 2010; Rodon and Levesque 2015; Stammler and Wilson 2006; Taylor et al. 2016; Wilson and Swiderska 2009; Petrov, 2010, 2012; Suopajärvi et al. 2016, including the use of different foresight techniques (Nilsson et al. 2015; van Oort et al. 2015; Karlsdottir et al. 2017).

In respect to determinants of sustainable development, SP1 outlined a large scope of work from modeling all aspects of Arctic economies, to a pan-Arctic comparison of sustainable development determinants, to detailed representations of Arctic economies and their dynamics. Although some progress was made in developing models of Arctic economies (e.g. Abele 2009; Berman et al., 2004; Petrov 2010; Huskey 2006; Lazhentsev 2005; Wenzel 2009; Winther et al. 2010), the gathering and analyzing of regional comparison data (Glomsrød, S., and Aslaksen 2006, 2008, Glomsrød et al. 2017; Petrov 2012), the examining of knowledge and other 'new' economic sectors (Petrov, 2008, 2016) and the conducting of in-depth regional case studies (e.g. Duhaime and Edouard 2015; Pelyasov et al. 2011; Zamyatina and Pelyasov 2015; Huskey 2006; Petrov 2010; Rasmussen 2014), relatively little has been done to synthesize understanding of subsistence and mixed economies outside specific case studies (Berman et al. 2004). As well, only limited interfacing of economic research with other disciplines was achieved.

SP1 also articulated the need to improve knowledge about regimes and institutions affecting sustainable development. This included a compilation of the different legal structures that influence large-scale and small-scale economies, an assessment of the relationships between various institutions and regimes, and an analysis of the role of property rights and local ownership, among others. In the last ten years considerable progress has been made on topics including co-management regimes (Berkes 2009; Kofinas 2009; Natcher 2013), Arctic Council studies (e.g. Koivurova and VanderZwaag 2007; Koivurova 2010; Axworthy et al. 2012; Kankaanpää and Young 2012; Nord 2016; Nilsson and Meek 2016; Heninen et al. 2016), UNCLOS and ocean governance (e.g. Jarashow et al. 2006; Carpenter 2009; Molenaar 2012c; Koivurova 2013) and other legal structures research (Loukacheva 2007; Shadian 2014). There has also been attention to the ongoing shift in legal status of indigenous peoples in both international law and national contexts (for review see Bankes and Koivurova 2013). The increasing recognition of indigenous peoples' rights has led to a shift in indigenous participation in various processes from "stakeholders" to "rightholders". Other studies tackled evolving property relationships in forestry (Lidestav et al. 2013, Tysiachniouk 2010), reindeer husbandry (Fondahl et. al. 2001; Forbes 2006) and changing fate control exercised by Arctic communities (Larsen et al. 2010), and different resource governance systems (e.g., Isaac and Knox 2004; Novikova 2016; Hovelsrud and Smit 2010). The examples listed here show that most studies are not focused on the economy per se, but rather discuss integrated livelihood-governance systems.

SP1 called for devoting significant attention to understanding Arctic demographics, health, urbanization and migration. Most of these suggested research themes demonstrated strong progress over the years. Many regional and circumpolar studies, including large multilateral interdisciplinary research and synthesis projects (e.g., BOREAS, MOVE, AHDR II) were completed (Huskey and Southcott 2010; Larsen, J. and Fondahl, G. 2014). We find substantial growth in the volume of research on Arctic health (e.g. Young and Bjerregard 2008; AMAP 2009; Chatwood et al. 2012; Young et al. 2013; ArcRisk 2013; Manchuk and Nadtochiy 2010), migration (e.g. Huskey and Southcott 2010; Ferris 2013; Laruelle 2016; Saxinger, 2016), and demography (e.g. Sokolova and Stepanov 2007; Carson 2011, Trovato and Romaniuk, A 2013; Axelson and Sköld 2011; Heleniak 2014).

Socio-economic costs of sustainable development and economic equity in the Arctic are the two remaining key areas identified in SP1. The Plan advocated rigorous assessments of trade-offs of sustainable economic development and focuses on the notion of "optimal economy" in the Arctic. Very little work on this topic has been completed since the release of the Science Plan. Perhaps one reason is that the concept of an "optimal economy" is incompatible with the regional and local variability of arctic economies. Even the broad idea of mixed arctic economies as combining land based activities, engagement with the cash sector and social relationships of sharing and cooperation is location specific, in the sense that species, employment dynamics and social structures differ by place. In this sense, there is not one mixed economy, there are many. Some progress, however, has been made, particularly in respect to assessing ecosystem services (Meltofte et al. 2013; Arctic Council 2015), investigating the tourism sector (Stewart et al. 2005; Grenier and Müller 2011; Hall and Saarinen 2010; Müller et al. 2013; Müller 2015), looking at impacts of non-resource development on single-industry and resource-dependent communities (Southcott 2009; Petrov, 2008). Mono-towns work in Russia (Dydik and Ryabova 2014; Zamyatina and Peliasov 2015), and evaluating scenarios associated with economic and climate change (Stephenson et al. 2013).

Understanding the relationship of equity and sustainable development was highlighted in SP1 and still remains a key research area as of now. The two overarching themes raised in SP1 were the analysis of the role of access to power and decision-making, and the impacts of sustainable development on equity, empowerment, and human development. Although considerable effort has been spent, only selected aspects of these key themes have been addressed, such as co-management practices (Bronen and Chapin 2013; Chapin III et al. 2009), stakeholder participation and incorporation of traditional knowledge in decision-making (Kruse et al. 2004; Meek 2011) and the role of equity in human development of Arctic residents (Larsen and Fondahl 2014). Most research was channelized through case studies, which have not yet been synthesized.

In summary, SP1 presented an ambitious agenda for research on various issues of sustainable development. Although some progress has been made on the themes it prioritized for research, most of the questions and themes remain relevant today. In addressing arctic economies and sustainable development, SP1 authors chose to focus on development issues from economic and social angles, and did not call for inclusion of other elements of sustainability science. SP2, 10 and 11 discussed below address some of these complementary perspectives.

3.2 ICARP II Science Plan 2. Indigenous peoples: Adaptation, adjustment, and empowerment

Science Plan 2 (SP2) begins by emphasizing that indigenous peoples are increasingly involved in research agendas as active participants and notes that research agendas set by Arctic indigenous peoples themselves, or reflecting indigenous cultures, will be a key factor in setting research priorities for next decade (p.1). It thus challenges the very sustainability of research involving indigenous communities and activities in the Arctic: it asserts the rights of indigenous peoples to shape research agendas and, in many cases, to participate in all stages of Arctic research. The plan then addresses four thematic issues: 1) Culture and Education; 2) Wellbeing and Health; 3) Economic Models; and 4) Indigenous Peoples and the State. As with our discussion of Science Plan 1, we summarize the key research priorities outlined in the Science Plan 2 that are related to sustainability and sustainable development, and briefly note progress achieved over the past decade. Throughout Science Plan 2, the issue of cultural sustainability is a "red thread," as is the issue of sustainable development of indigenous communities.

In terms of Culture and Education, key research priorities posed by SP2 include the success of language revitalization programs and the criticality of recording oral history and traditional knowledge for cultural sustainability. Over the past decade, significant work has been carried out on linguistic continuity, loss and revitalization (Barry et al. 2013; Grenoble and Olsen 2014; Schweitzer et al. 2014). Two Arctic indigenous language symposiums, organized by the Inuit Circumpolar Council in 2008 and 2015, addressed current status, metrics, and best practices for revitalization (ICC Canada 2012). The Arctic Council's Arctic Biodiversity Assessment project included a focus on linguistic retention issues (Barry et al. 2013). Substantial work has also been carried out on oral history and traditional knowledge in both the North American and Eurasian North (e.g. Berkes 2012; Collignon 2006; Crate 2006; Cruikshank 2005; Dudeck 2013; Gearheard et al. 2011; Krupnik and Jolly 2002; Krupnik et al. 2010; Riseth et al. 2011; Vlassova 2006; Lavrillier 2013; Filonchik 2011; Vinogradova 2005; Wilson and Sviderska 2008; Mustonen and Mustonen 2016). Science Plan 2 also raises the issue of how indigenous knowledge is being used to inform western scientific research (especially on climate change and wildlife management). Several major projects can be identified as leading examples in working toward such linking (Gearheard et al. 2010; Eira et al. 2013). We see less progress in more holistic and synthetic work, called for by SP2, in inter-rogating "the interplay of historical, sociological, political and economic factors that explain why some indigenous groups are perceived to be strong, others weak, and investigating the related role of public policies in promot-ing indigenous cultural and linguistic continuity" (Dahl et al. 2005, 6).

SP2 also asks a set of questions about education related to sustainable development, including best practices in educational programs, the role of indigenous cultural material in education programs, challenges posed by bilingualism, and the effect on local cultural well-being of moving away from humanities subjects (arts, creative writing, music, art, sports, etc.) toward vocational and industrial training. Research on incorporating lan-guage and traditional knowledge in curriculum, and more generally, on 'decolonizing' indigenous education as a step toward cultural sustainabil-ity, has advanced in the last decade (Aylward 2009, 2010,; Barnhardt 2005, 2014; McGregor et al. 2010, 2012; Patrick 2005; Nabok 2014). In Russia, the introduction of nomadic schools has been examined as a "best prac-tice" in providing for cultural continuity (Lavrillier 2013). Challenges to school completion have also been investigated (Berger and Epp 2006). Inuit Tapiriit Kanatami, representing the Inuit of Nunavut, Nunatsiavut, Nunavik and the Inuvialuit region of the NWT, published a National Strategy on Inuit Education (ITK 2011) that addresses several of the above issues.

In terms of the Health and Well-Being thematic area, a key question posed by SP2 is "What kinds of control – economic, social, political, and cultural – are essential for the well-being of individuals and communities?" (Dahl et al. 2005, 10). This question strikes at the heart of "fate control" as a key aspect of sustainable communities. Ongoing analysis of the SLiCA (Survey of Living Conditions in the Arctic) data sheds light on various aspects of well-being (e.g. Poppel et al. 2007; Poppel and Kruse 2009; see also Rautio et al. 2014 and Brustad et al. 2014). Arctic Social Indicators Report (Larsen et al. 2010) identified "fate control" as one of the six key domains of human well-being in the Arctic (Dahl et al. 2010). SP2 poses more specific questions about family resilience (and lack thereof), well-being during pregnancy and early childhood and their effect on lifelong well-being, an area that still has received only limited attention (but see Allen et al. 2014; Bals et al. 2011; Salokangas and Parlee 2009). Some progress has been made on examining issues of violence, sexual abuse, and struc-tural racism (Hansen 2010; Juutilainen et al. 2014, Sunnari 2010) as well as suicide (e.g. Larsen, C.V.L. et al. 2010; Larsen et al. 2010; Poppel 2006).

The importance of traditional food for physical well-being, while not the focus of a specific question in SP2 was underscored repeatedly in its text. Research on food security and dietary issues in the Arctic has progressed in the past decade (e.g. Chan et al. 2006; Bjerregaard and Mulvad 2012; Ford et al. 2011; Dudarev 2012; Dudarev et al. 2013; Huet et al. 2012; Loring and Gerlach 2009; Nilsson L.M. et al. 2013; ICC Alaska 2015). Also not called for explicitly by SP2, but seeing substantial development is research on climate change and indigenous health, both physical and mental, in the Arctic (Cunsolo Willox et al. 2012, 2013a, 2013b; Greer et al. 2008).

The words "sustainability", "sustainable development " and "sustainable use" appear repeatedly in SP2's third section/theme area on Economic Models. Research questions posed included implications of barriers, legal and regulatory to local indigenous business interests (for some examples of work addressing this, see Dana et al. 2008; Wilson and Alcantara 2012; Southcott, 2015); the social and cultural implications of moving from common use rights to private property rights and the significance of various kinds of "ownership" (e.g. Carothers et al. 2010, Novikova 2010; Tishkov et al. 2015). Emerging governance approaches to renewable resource use, and their ability to promote sustainable use and to empower indigenous peoples was posed as an area needing research, as was exploring whether co-management and governance institutions increase resilience. Both these questions have received significant attention in the past decade (e.g. Coates and Poelzer 2010; Forbes 2013; and specifically on co-management, Armitage et al. 2008, 2010; Berkes 2009; Kofinas 2009; Natcher 2013). In terms of non-renewable resource management, SP2 asks whether decisions regarding non-renewable resource development meet international norms requiring free, prior and informed consent. The past decade has witnessed the development of research on community engagement regarding non-renewable resources (see Aaen 2012; Stammler and Peskov 2008; Stammler and Wilson 2006).

"Indigenous peoples and the State" is the final theme area suggested under SP2. Research questions related to cultural and social sustainability include how legal pluralism promote protection of indigenous rights and how traditional indigenous legal knowledge can be "translated" into the language of official jurisprudence (Dahl et al. 2005, 14) Legal pluralism has received some attention, though mostly in the Fennoscandian context (e.g. Svensson 2005; Watson Hamilton 2013). There has also been research attention to the potential of a Saami Convention for protecting Saami rights across states (Koivurova 2008a) SP2 also asked to what extent international law and international processes influence the life of indigenous peoples (Dahl et al. 2005, 15), and how successful has the Arctic Council been in building capacity of indigenous people as participants in its activities (ibid).

Researchers documented the increased role of indigenous people and organizations in international governance processes (e.g. Abele and Rodon 2007; Heinämäki 2009; Loukacheva 2009), including specifically in the Arctic Council (Koivurova 2010). In terms of international law's influence on indigenous northerners, we see a burgeoning of research over the past decade (e.g. Bankes 2011; Bankes and Koivurova 2013; Koivurova and Stepian 2011; Loukacheva 2010). This includes the studies of environmental "refugees" from relocated communities (Bronen 2011).

SP2 ends by underscoring again the importance of indigenous peoples owning and actively participating in research activities. Contributions from a growing number of indigenous researchers from the Arctic (e.g. Brattland 2014; Metcalf and Robards 2008; Oskal et al. 2009; Rasmus 2008; Ulturgasheva 2012), and increased research partnerships between southern researchers and northern community members (e.g. Ulturgasheva et al. 2011; Gordon, 2015) indicate that progress is being made on this front, not last in connection with the International Polar Year 2007–2008 (Krupnik et al. 2011); see the Methodology section for further discussion on community-based research in the Arctic.

3.3 ICARP II Science Plan 10. Rapid change, resilience and vulnerability of social–ecological systems of the Arctic

Science plan 10 (SP10: Kofinas et al. 2005) was written when research using resilience as a framework was just starting to take off. One of the main messages in the science plan was a push to use social–ecological systems as the unit of analysis, implying further attention to the links between social and ecological processes. The plan also highlights a focus on the dynamics of complex systems that had been developed within resilience research, including attention to thresholds and irreversibility. The title also mirrors a scientific discourse in which vulnerability was a prominent theme, as discussed in Chapter 4), and the vulnerability and resilience frameworks were starting to be in dialogue with each other. Another example of this synthesis, aside from SP10, is chapter 17 of the Arctic Climate Impact Assessment: Climate Change in the Context of Multiple Stressors and Resilience (McCarthy et al. 2005).

Since then, there have been a growing number of publications referring to both the Arctic and resilience, as well as research projects that apply a resilience-inspired analytical framework. Examples include studies of reindeer herding in northern Norway (Mathiesen et al. 2013) and Yamal (Forbes et al. 2009), transformation of subsistence systems in Canada (Clark and Workman 2013), community resilience among youth as it relates to the ecosystem (Broderstad and Eythórsson 2014; Nystad et al. 2014) as well as several studies of social–ecological systems in Alaska (Berman et al. 2017; Kofinas et al. 2010), which is also mentioned in Chapter 4 of this book. The resilience framing has started to enter the policy arena, most notably with the Arctic Council project Arctic Resilience Report, including the publication of an interim report in 2013 (Arctic Council 2013, 2016).

While it is easy to conclude that resilience-thinking has become established as a useful framework for characterizing patterns of social–ecological change in the Arctic, the results from a literature search also reveals that the word *resilience* is often used in contexts where the social–ecological system is not the analytical unit. This includes discipline-focused ecological studies as well as studies of human well-being undertaken at the household and community scales, and mental well-being focused at the level of individuals. In fact, only a small proportion of the publications that use the word "resilience" explicitly focus on the links between ecological processes and social processes beyond the superficial level.

With this caveat, major progress has nevertheless been made in relation to some of the specific issues that are mentioned as priority research themes since SP10. This includes increasing attention to potential threshold changes in the Arctic, sometimes framed as tipping points. Two prominent examples are the special issue of *Ambio* on Arctic tipping points, which was the result of a large EU project (Wassmann and Lenton 2012) and the Arctic Resilience Report (Arctic Council 2016). Thresholds are also in focus for one of the chapter in the Arctic Resilience Interim Report 2013 (Cornell et al. 2013; Peterson and Rocha 2016). A common lesson from these publications is that the notion of thresholds or tipping points appears to be more relevant for the physical sciences related to climate change (for example Arctic sea ice as a potential tipping point in the global climate system), than for the social science where the notion has been criticized because it does not include attention to agency and power (Nuttall 2012; Nilsson and Koivurova 2016). A focus on tipping points has similarities to social science concepts such as path dependency, but the relevant social science literature is not as likely to mention resilience.

Another major area of progress relates to the SP10 question: "What are the factors that account for variance in systems and subsystems?" Progress here stems mainly from research on vulnerability and adaptation, where adaptation in particular has come into focus much more prominently in research over the past decade. This includes a growing body of local case studies, some of which have been part of larger circumpolar efforts such as CAVIAR (Hovelsrud and Smit 2010), ELOKA (ELOKA 2015; Pulsifer et al. 2012) and SIKU (Krupnik et al. 2010). While some/ most of this research has been motivated by the need to better understand local impacts and responses to climate change, the result has been a more nuanced understanding of relevant social complexities in the local context than was usual in early impact studies. It includes identifying determinants of adaptive capacity (e.g. Smit and Pilifosova 2001; Wesche and Armitage 2010, Vulturius and Keskitalo 2013; Kofinas et al. 2013) as well as an emerging understanding that adaptive capacity in itself is not sufficient if it is not activated. Kofinas et al. (2013) explicitly links this discussion to a resilience framing.

SP10 highlights a need to identify indicators of vulnerability and resilience. While work on indicators for human development in the Arctic has developed substantially (see Chapter 4), thinking about resilience indicators is much more recent. An example is the WWF RACER project, which looks at potential indicators for ecological resilience, but without attention to social processes (Christie and Sommerkorn 2012). There is thus a need for further work on resilience indicators that take both social and ecological dynamics into account.

While further work on indicators should be informed by progress on understanding attributes of systems that provide resilience, there is also a continued need to gather and synthesize data across existing and new case studies in a systematic manner. SP10 suggests a meta-database using a standardized format and a common set of key variables. Some such work is presented in the Arctic Resilience Report (Arctic Council 2016), but more efforts are needed to meet some of the ongoing challenges in synthesis efforts. This includes a requirement for both strong contextual understanding of each case study and the willingness and time for researchers to participate in synthesis efforts that may require different approaches and analytical frameworks than those that motivated the initial research.

The last key research question identified in SP10 focuses on how the study of resilience and vulnerability should inform policy. This includes breaking down the science-policy barrier in ways that promote learning and co-production of knowledge, but also the development and application of linked social/ecological models as decision-support tools that illustrate key change mechanisms and potential interactions between ecological, human well-being and economic outcomes within arctic systems. From the perspective of integrated assessment, the Arctic Community Synthesis model was developed and applied in Alaskan and Canadian communities (Berman et al. 2004; Kruse et al. 2004), and an adaptive management framework was developed to assess cumulative effects of climate change and industrial development on caribou populations (Gunn et al. 2014). Agent-based modeling is ongoing by around emerging tradeoffs in mixed economies in Alaska (BurnSilver et al. 2017). Such modeling efforts can be used as boundary objects for co-learning, which might be especially relevant in local and policy contexts. Other on-going efforts include developing locally relevant narrative scenarios that are also informed by the global context and using these as part of the knowledge base in circumpolar assessment processes, such as the Arctic Council project Adaptation Action for a Changing Arctic (AACA) (Baggio et al. 2016; Nilsson et al. 2015; Van Oort et al. 2015). These efforts also link to a call for research methods that facilitate co-learning (Paschen and Ison 2014). At the international level co-learning processes are more challenging, especially if they relate to conflicting political interests or different views of the role of science in relation to policy. Here there is a need for experimentation with different models for communication at the science-policy interface in ways that include attention to factors that affect credibility of the knowledge, legitimacy of the process and salience of the questions asked for the policy makers (Mitchell et al. 2006). Credibility, for example, is affected by what knowledge traditions are included in the analysis and by how independent the process is from strong political influence. Regarding salience, a general observation is that linking resilience to adaptation has provided a means to make research more policy-relevant and increase the contribution of empirical work to theory. However, there are also challenges. Many of the drivers of change and decisions that affect resilience of social–ecological systems are linked to political interests, where social science is likely to provide insights that may not be politically convenient. SP10 describes the need to pose questions such as "resilience for whom and of what," not least in relation to decisions (or non-decisions) that may lead to transformative change that benefit some actors but not others. These equity dimensions of resilience, together with attention to agency and power, were not very prominent in 2005 but have become increasingly important in recent years (e.g. Arctic Council 2013, 2016). We foresee the need for continued work that includes both the focus on system dynamics provided by the resilience lens and attention to agency, power and inequality.

3.4 ICARP II Science Plan 11. Arctic science in the public interest

The goal of Science Plan 11 (SP11) "Arctic Science in the Public Interests" was to "understand better the driving forces behind Arctic science and its relationship to the public interest," Young et al. 2005, 1) The focus was on 1) understanding the image of the Arctic and Arctic science 2) the construction of research questions 3) the conduct of research in the Arctic 4) the control and communication of knowledge, and 5) the impacts and relevance of research. It mirrored a growing research attention to the relationship between science and society in humanities and social science research as well as the increasing articulation of indigenous concerns about the impacts of research, as discussed for SP2. When it was written for ICARP II, the Arctic as a region was just starting to attract global interest. The results of the Arctic Climate Impact Assessment had become public and focused the world's attention on the impacts of on-going climate change as something happening here and now (ACIA 2004; ACIA 2005). However, SP11 pre-dated the real game changer in term of Arctic sustainability becoming part and parcel of general public interest: the sea-ice minimum of 2007 (Christensen et al. 2013). This unexpected "event" became a news item that travelled the world and fundamentally changed not only images of the Arctic but also perceptions of public interest. Issues related to military security, new maritime transport routes and resources exploitation featured in major newspapers and other media outlets, as well as in both academic literature and popular science books (e.g. Emmerson 2010; Breum 2011). Moreover, an increasing numbers of actors started to debate the need for new political initiatives or reforms in relation to governance of the region in light of the new turbulent interests (e.g. Koivurova and VanderZwaag 2007; Arctic Governance Project 2010; Koivurova 2008b; Young 2011).

This new wave of attention to the Arctic region, which persists, makes the need to understand the role of science, and knowledge production more generally, even more urgent today than 10 years ago. Moreover it makes many of the issues raised in the ICARP II science plan relevant in light of an increasing number of actors and interests: How do we understand the images of the Arctic and Arctic science and the forces shaping those images? How is research about the Arctic shaped and priorities set? How is it influenced by international politics? Other questions have emerged because of the increasing volume and thus potential impacts of scientific activities by external actors in the Arctic. They include issues related to impacts of research, such as the role of ethical guidelines and participatory research practices, and to the control and communication of knowledge, including the role of traditional knowledge. Increasing attention from actors without knowledge about the region also makes it more important to highlight the internal diversity across the Arctic and the various local perspectives that have been silenced in many narrations of the region. Below we provide a few examples of increasing scientific activity in the areas suggested in the ICARP II Science Plan 11, without any ambition to be comprehensive.

The focus on understanding the production of images of the Arctic highlights both externally and internally produced images, how they have changed over time and how they affect research. SP11 highlights the history, sociology and politics of science as a priority approach, and these are probably among the best developed of the highlighted research areas. It represents a continuation of research that was underway (e.g. Bravo and Sörlin 2002; Steinberg et al. 2015), focusing on the intersection between science and historical narrations of the Arctic, also including attention to themes such as nation building and geopolitics (e.g. Sörlin 2013; Heymann et al. 2010). This focus has also been applied to contemporary dynamics at the science-politics interface (e.g. Christensen et al. 2013). Moreover, in connection with the International Polar Year, the history and politics of IPY as such also came into focus (e.g. Shadian and Tennberg 2009; Elzinga 2013). Most of this work appears to address large-scale processes rather than the local sociological and political dynamics of research. There are many reasons to pay more attention also to the national and local sociologies and histories of science, which would provide an important knowledge foundation for understanding the legacies that still affect research practice in the region. This theme links to the issue of research practice and conduct, another major theme in SP11. SP11 also highlights a need to identify and understand the production of "internal images of the Arctic" building further on ethnographic methods and with attention to the diversity of communities across the Arctic. In relation to indigenous peoples such research has a long tradition with continued work in the past decade. The relationship between modernity and traditions highlighted in AHDR II (Larsen and Fondahl 2014) is a key issue of relevance for understanding the construction of contemporary images, especially among young people. A new development is increasing research on the political identity of indigenous peoples (Lantto 2013; Shadian 2014).

The second theme in SP11 is the construction of research questions. It relates to overall images discussed above but is more specific in its focus, including a question of the actors behind processes such as IPY and ICARP. While there has been some research attention to IPY as previously mentioned and also to new scientific actors, including scientific attention from non-arctic countries (Paglia 2016), major work remains to be done to trace who defines the priorities of Arctic research.

Regarding theme 3 – the conduct of Arctic research, there is increasing discussion, including attention to participatory methods and process to ensure that research efforts address community needs (see discussion and references in Chapter 5). Moreover, the IPY of 2007/2008 moved positions forward considerably regarding the inclusion of traditional knowledge in polar research and indigenous peoples as active participants in the research process (Krupnik et al. 2011; Krupnik 2010), which is also becoming the norm in scientific assessment under the auspices of the Arctic Council. However, the surveys of how ethical guidelines are used and the impacts of research that were suggested in SP11 have not materialized (with a few exceptions, cf. Gordon 2015) and there is thus a risk that new norms are formulated on paper without necessarily being put into practice, or that good intentions have impacts that are neither intended nor desirable for those who will be affected by the research. In a similar vein it would be relevant to more systematically follow up on the outcome of different efforts to include traditional knowledge in research projects and assessments.

A development that was not discussed in SP11 is the increasing role of commercial actors in the Arctic and the connections between production of new knowledge and economic (and political) interests, either directly such as geological surveys related to mineral and hydrocarbon resources or indirectly by using these activities as a way to claim stakeholdership in the Arctic (Konyshev and Sergunin 2012; Laruelle 2013; Steinberg et al. 2015; Yagiva et al 2015; Paglia 2016). A challenge for the ethical conduct of research is the lack of knowledge and other priorities among research actors without previous history in the region. Moreover, with the current discourse of natural resources-based economic development, there is an increasing risk that national economic and security interests supersede local priorities. A key research priority would thus be to study the impact and conduct of new research actors and interests in the Arctic. Noting that an earlier wave of commercial interest in the Arctic – the oil boom of the 1970s – played an instrumental role in increasing political awareness and organizational capacity among indigenous peoples (Shadian 2014), and by extension has also affected how researchers have been forced to pay attention to indigenous peoples' perspectives. Such research would also be closely related to the fourth and fifth theme in SP11 (i.e. control and communication of research and impacts of research respectively). In theme 4, SP11 highlights that knowledge is a form of empowerment and that lack of control of knowledge is therefore a form of disempowerment and lists a range of questions that should be addressed in research. While some of the issues are implicitly included in some of the studies of history and sociology of science cited above, most of them have yet to be explicitly and systematically addressed and are still relevant for the coming decade. Examples include: How are research findings used in decision-making? Which forms of knowledge are seen as legitimate? Who controls the economic benefits of knowledge developed in the Arctic?

A recently emerging theme is Foucauldian-inspired studies of gender and indigeneity that specifically address knowledge as power to shape images of the Arctic. Examples include studies that highlight a tendency to essentialize indigenous peoples and limit their knowledge contribution to issues centering around their relationship with nature (Lindroth and Sinevaara-Niskanen 2013), while gender issues mainly considered in relation to community and family but not to the broader economic structures (Sinevaara-Niskanen 2013; but see Oddsdóttir et al. 2015). Regarding theme 5 – the Impact of Research – there has been some attention in the past decade to the policy impact, or lack of such impacts) from environmental assessment under the auspices of the Arctic Council (e.g. Nilsson 2012; Kankaanpää and Young 2012; Arctic Council 2016), but less attention to the broader perspectives, and much work thus remains to be done in this area.

In summary, the issues raised in SP11 are more relevant than ever due to the growing attention on the region. There is an increasing body of research about the history, sociology, and politics of Arctic science, which has the potential to move forward our understanding of the role of science in society substantially in the next ten years (Shadian 2009; Sörlin 2013). There is also increasing interest in the conduct of ethically sound and inclusive research in general. There remains a need, however, for systematic evaluation and critical study of consequences of both old and new practices, including attention to the role of new commercial actors and to the intersection between national and local priorities.

4 Progress in Arctic sustainability research 1

Theoretical developments in Arctic sustainability science

4.1 Progress and milestones

As described earlier, the sustainability concept has deep roots that recognize the importance of sustainable interactions between people and ecosystems if future generations are to thrive. Three broad sets of concerns came together at the end of the 20th century to raise the profile of sustainability as a practical and scientific issue. 1) In developed nations, rising awareness of environmental degradation in the 1960s and 1970s led to an environmental movement that sought to reduce environmental threats. 2) At the international level, there were rising concerns in the 1980s about the need to support economic development and poverty alleviation in developing nations, while sustaining the environment and natural resources to support the needs of future generations (the Brundtland Report; WCED 1987). 3) Finally within the scientific community, in the 1990s and 2000s there was rising recognition of the global scale of environmental degradation and the inadequacy of current scientific approaches, leading to formulation of the need for "sustainability science" to address the long-term interactions of people and nature (Clark 2007; Kates et al. 2001; MEA 2005; NRC 1999). Sustainability science is an interdisciplinary science that addresses the long-term interactions of people and nature. Like agricultural and health sciences, sustainability science is problem-focused, seeking to enhance fundamental understanding in order to solve problems (Clark 2007; Stokes 1997). In the global aggregate, 60% of the benefits that people derive from ecosystems (ecosystem services) are declining, and only the three most intensively managed services (crop production, aquaculture, and livestock) have increased (MEA 2005). Because of the strong ties between global change research and the emergence of sustainability science, vulnerability and the understanding of societal risks from global change thus became the central framework for sustainability science (Turner et al. 2003).

While the Arctic was not a focus of the initial reports of the International Panel for Climate Change (IPCC), it received increasing attention in national, international, and Arctic assessments as focus shifted to regional impacts of climate change in the late 1990s, and the Arctic was identified as a particularly vulnerable region (IPCC 1997; ACIA 2005; IPCC 2001; IPCC 2014; McCarthy et al. 2005; Nilsson and Döscher 2013). Earlier concerns about global and local impacts of persistent organic pollutants and heavy metals and their accumulation in animals and people also highlighted the vulnerability of the Arctic regions to and their dependence on global processes (AMAP 1998; AMAP 2002). Thus, the emergence of global change science and sustainability science, both based on concerns about global trends, coincided with the ICARP II planning process, which was intended to address arctic dimensions of global environmental change. Given the strong role of vulnerability analysis in both IPCC and sustainability science, it is not surprising that this became the principal framing for sustainability science in ICARP II, as discussed below.

From the beginning, sustainability science took a systems focus. It viewed systems as self-organized complex adaptive systems – i.e. systems that respond to change through changes in not only their fundamental properties but also the nature of their responses to subsequent changes (Levin 1998). There has been considerable speculation about how complex changes in physical components of the Arctic system might lead to non-linear changes or transformations, for example through loss of sea ice, or changes in deep-water formation and circulation of the global oceans (Schellnhuber 2004). Changes in the social–ecological system have also been discussed in relation to radical changes in governance (Young 2012; Nilsson and Koivurova 2016) and to the interactions between people and nature (Duarte and Wassmann 2011).

The concept of complex adaptive systems is consistent with observations from the planning literature that recognized the "wicked" nature of social–ecological problems (Rittel and Webber 1973). In other words, every partial solution to a planning problem creates new problems or frames the problem in a new context. Consequently, complex problems can never be truly "solved." The challenge is to build resilience that reduces the likelihood of undesirable changes (Chapin et al. 2006; Tishkov 2012) or facilitates transformations to escape undesirable situations, such as poverty traps (Kofinas et al. 2010).

Thus it is essential that sustainability science acknowledges uncertainty in both drivers and outcomes. This approach is well developed in Arctic vulnerability studies discussed below. For example, suppression of wildfires, which have increased in frequency due to warming has maintained stands of flammable forests near communities, thereby increasing their vulnerability to future wildfires (Chapin III et al. 2008). Another trajectory has been to embrace the processes of co-learning and institution building as elements of sustainability (Reed 2008). The wicked problems that characterize complex adaptive systems have led sustainability science to focus at various times on either building resilience or flexibility to future conditions, or decreasing vulnerability to current conditions, which presents a conceptual tension in the literature that will be explored below.

In general Arctic sustainability research demonstrates quickly increasing volume (Figures 4.1 and 4.2) and growing theoretical and methodological strengths. Initial system science in the Arctic focused largely on biophysical processes and human impacts on these processes but failed to systematically address the coupled linkages between people and nature. Research at Cape Thompson, Alaska (one of the first ecosystem studies; Wilimovsky and Wolfe 1966), for example, considered individual ecosystem components, including plants, animals, and people, but did not address their coupled interactions. Similarly, the Tundra Biome Programme of the International Biological Programme, contributed substantially to integrated ecosystem research but largely ignored coupling to the human component (Bliss et al. 1981; Brawn et al. 1980). Interactions between ecosystems and society largely focused on human impacts (Walker et al. 1987). However, coupling of the human dimension with biophysical studies was substantially developed through the Arctic Council's pollution and climate assessments (ACIA 2005; AMAP 1998). A coupled system focus that draws explicitly on sustainability science has been fostered by various national and international funding programs (for example, the "Arctic Observing and Research for Sustainability" Collaborative Research Action of the Belmont Forum, ArcSEES, International Social Science Council) (Chapin et al. 2006; NRC 2003).

Progress toward understanding coupled SESs is represented by Chapters 3 and 12 in ACIA (Huntington et al. 2005; Nuttall et al. 2005), in which a long tradition of arctic anthropological research demonstrating strong linkages between indigenous peoples and their environment(s) (Nelson 1973; Nelson 1983) was connected to traditional ecological knowledge (TEK) and resilience. This work emphasized TEK as representative of a different kind of human-nature knowledge, an understanding of which could contribute to more nuanced conceptualizations of coupled human-ecosystem interactions. Berkes et al. (2000, 1251) described this knowledge as consisting of fine-scale, long-term, and place-based observations that manifest in the rituals and practices of the everyday lives of indigenous peoples. For example, TEK describes diverse human-animal-landscape relationships that frame

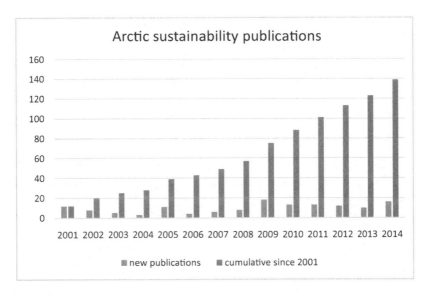

Figure 4.1 Literature on Arctic sustainability 2001–2014

Source: Google Scholar, "Arctic" and "sustainable," titles only

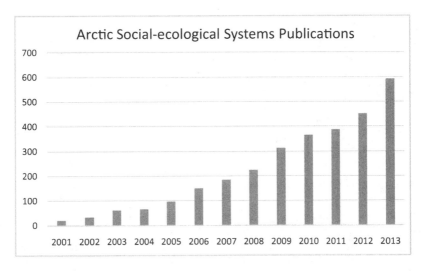

Figure 4.2 Literature on Arctic social–ecological systems 2001–2013

Source: Google Scholar, all text

human sharing and cooperation as well as broader institutional arrangements (Wenzel 2004). A challenge has been to integrate scientific and indigenous epistemologies in ways that mutually inform understandings of coupled systems, but take care to not use Western science to "validate" TEK (Nadasdy 1999) or ignore power differences often inherent in these very different modes of knowing, understanding and communicating. Berkes et al. (2000) identified a range of indigenous strategies for interacting with ecosystems that incorporate surprise, monitor change, and manage for ecological heterogeneity, all actions in line with complex adaptive systems thinking. Strongly collaborative efforts between scientists and local hunters/observers are now ongoing in multiple contexts to document knowledge about ecological dynamics and climate change (Huntington et al. 2005; Gearheard et al. 2006; Gearheard et al. 2010; Cochran et al. 2013; Poddubikov 2012; Vlassova 2006). These efforts to understand coupled human-environment interactions from multiple perspectives are important and timely, given the pace and extent of arctic change (Jorgensen 1990; Ford et al. 2012; Krupnik and Jolly 2002).

4.2 Vulnerability, resilience, and sustainability

Vulnerability and resilience have become major bridging concepts within sustainability science. The frameworks emerged from different social and biophysical disciplines, but they currently provide a primary means for conceptualizing and operationalizing complex systems thinking in coupled social–ecological systems. There remain grey areas of conceptual overlap between key concepts endemic to both frameworks, for example, adaptation, adaptability, adaptive capacity, and transformation. Application of these frameworks to date in the Arctic has been propelled forward based on climate as a primary exposure driving change, but both approaches are directly relevant to consideration of the broader array of economic, political, and institutional drivers framing questions of future sustainability in the Arctic.

Vulnerability. In the IPCC Third Assessment Report, climate change vulnerability was defined as a function of exposure, sensitivity and adaptive capacity (McCarthy et al. 2001). Adger (2006, 268) more specifically defined vulnerability as "the susceptibility to harm from exposure to stresses associated with environmental and social change and from the absence of capacity to adapt." Between these two definitions of vulnerability lies a range of assumptions, disciplinary context, policy goals and applications. In an effort to clarify the use of vulnerability concepts, O'Brien et al. (2004) differentiated the study of vulnerability into two broad domains. The first considered vulnerability as an "endpoint" (Kelly and Adger 2000, 326), whereby the propensity for harm (vulnerability) was understood as the remaining impact of change after adaptation had occurred (O'Brien et al. 2004). The second domain considered vulnerability to be the starting point from which underlying patterns of drivers, sensitivity, and adaptive capacity could be disaggregated and assessed. This second domain focused on defining who was likely to be affected by exposures under current conditions – and, what these patterns implied for future adaptability. Both domains are represented within arctic vulnerability research, but they emerge from different disciplinary traditions and focus on different mechanisms of vulnerability, time scales of adaptation and ensuing implications for sustainability.

The field of vulnerability studies leveraged multiple research traditions, among them hazards and disaster research (Burton et al. 1978; Blaikie et al. 1994), the sustainable livelihoods framework and political economy/ecology (Ellis 2000, Eakin and Luers 2006). Differences between the two vulnerability domains described above hinge on the degree to which biophysical drivers are prioritized (end point focus) vs. integrated with social concerns (starting point focus). Prioritizing ecological indicators assumes that those living in the riskiest environments are the most vulnerable (Liverman 2001). This approach has framed climate vulnerability assessments, first using predictions for future emissions to develop climate scenarios that are subsequently overlaid onto landscapes to predict impacts (IPCC 2014). The focus is then to identify adaptation options that would be effective in reducing impacts of climate change, (i.e. Impact – Adaptation = Vulnerability) (Burton et al. 2002). Climate vulnerability assessments in this tradition have been undertaken for all Arctic countries (Ford et al. 2010; Melillo et al. 2014; Miljoverndepartement 1991; Klimat och sårbarhetsutredningen 1997; Sharakhmatova 2011; Sulyandziga and Vlassova 2001), under which different options for adaptation are evaluated given the "known" capabilities of sectors, economies or specific ecosystems. Adaptation options in this domain have tended to focus on technological or infrastructure changes that would decrease impact under predicted future conditions (O'Brien 2004). For example, against predicted impacts of thawing permafrost and coastal erosion, the options of relocating communities versus infrastructure replacement/repair without relocation are evaluated (Chapin III et al. 2014).

In contrast, the second research domain of vulnerability research integrates multiple factors – political, social, economic and ecological – which interact to generate a state of vulnerability for livelihoods or specific social groups (households, communities, etc.). This state of vulnerability is expected to set the stage for future adaptation. This contrasts with the role adaptation plays in the end point view of vulnerability, in which vulnerability is a result of prior adaptation. The sustainable livelihoods approach has been used to elucidate the social side of the vulnerability within coupled systems. The approach posited that that groups or households are characterized by a diverse range of asset types or entitlements (i.e. natural, social, human, financial, physical/ infrastructure) (Sen 1981; Scoones 1998). In combination, these assets define who would be exposed to poverty, and thus how vulnerable are people and livelihoods. Political economy/ecology frameworks additionally contributed a nuanced consideration of how power and equity dynamics affect poverty outcomes (Kelly and Adger 2000). Thus, households are vulnerable not only based on ecological hazards, but also as a function of the relative power and equity dynamics that both define access to key resources and locate people in places with higher risk to ecological hazards (Adger 2006).

Vulnerability here is an outcome of processes that open or constrain options for people. Interactions across concept areas are significant: 1) exposures, which have specific patterns (e.g. frequency, duration, timing), 2) sensitivity, defining the level of response expected given specific patterns of user interactions with resources, and 3) adaptive capacity, or the set of resources or endowments available to actors and the ability to employ them, which are pre-conditions for adaptation (Adger 2000). Questions posed within this second vulnerability domain are: given particular combinations of exposures, what are the processes and capabilities that increase sensitivity (a negative) or enhance adaptive capacity (a positive)? The approach facilitates consideration of why some people or groups are more likely to be affected than others by specific exposures.

Two trajectories of vulnerability research have emerged from this second domain (i.e. vulnerability as a starting point). The distinction hinges on questions of scale (community vs. other scales of analysis) and the relative emphasis placed on future adaptation based on identified adaptive capacity. The first trajectory identifies the relative vulnerability of groups or livelihoods based on indicators of sustainability or adaptive capacity that are weighted empirically in some way. Vulnerability analyses of this kind are more numerous outside the Arctic context (Eakin and Bojorquez-Tapia 2008; Nelson et al. 2010; Yohe and Tol 2002). A key strength of this approach is that results highlight the range of social and economic capabilities and sensitivities within groups that result in heterogeneous vulnerability outcomes under change (Adger 2006; O'Brien 2004).

4.3 Vulnerability assessment

The second trajectory is also framed as vulnerability assessment, but the goal is to identify relevant exposure-sensitivities and adaptive capacity from community perspectives, which contribute to qualitative understandings of

how change is experienced. The goal here is to make practical contributions to decision-making (Smit and Wandel 2006). A series of papers has
described the vulnerability of Canadian indigenous communities to climate
change from this perspective, where engagement is explicitly oriented to
describe community-level sources of adaptive capacity (i.e. local knowledge and actions) that may be key sources of adaptation (Ford and Smit
2004; Ford et al. 2006; Ford et al. 2008; Ford and Furgal 2009; Ford and
Pearce 2010; Ford et al. 2010). Under the auspices of the CAVIAR project
(Smit et al. 2008; Hovelsrud and Smit 2010) this assessment framework
has been scaled up to make comparisons of climate change vulnerability
based on case studies of communities of different sizes (urban/rural), levels
of remoteness and livelihood types (Smit et al. 2008; West and Hovelsrud
2008; Hovelsrud and Smit 2010). Findings suggest that once exposure-
sensitivities are identified, governance and institutional capacity to set
and implement priorities emerges as a key lever for community adaptation
(Dannevig et al. 2012).

As should be clear from above, the vulnerability framework is con-
ceptually very rich, and has been applied to a diversity of contexts and
questions in the Arctic and elsewhere. This diversity has led to criticisms
that the concept is too amorphous (Gallopin 2006). Climate change has
been the dominant exposure of interest in Arctic vulnerability studies to
date. However, Kofinas et al. (2000) emphasize that climate is only one of a
number of important drivers of change of current concern in the Arctic and
these will have cumulative effects on sustainable outcomes for people and
landscapes. There are also tradeoffs associated with different approaches to
vulnerability assessment from a methodological perspective. Application of
diverse definitions and metrics (qualitative vs. quantitative), scales of analysis (household vs. community, region, etc.) have strengths and weaknesses
given particular goals (Berman et al. 2017). This diversity of approaches
has made spatial and temporal comparisons of vulnerability outcomes and
dynamics across arctic contexts difficult to achieve.

4.4 Resilience

Resilience has emerged as another integrative concept in sustainability
science and research. It is defined as the capacity of a system to absorb
disturbance and re-organize so as to maintain essential identity, structures,
dynamics, and the capacity for adaptation, learning and transformation
(Walker et al. 2004; Walker and Salt 2006; Folke et al. 2006). Resilience
theory posits four interrelated system functions in an adaptive cycle meta-
phor: release, reorganization, growth, and conservation (Holling 2002;
Ziker 2012; Arctic Council 2013). Each of these functions varies according

to the accumulation of capital (active to stored) and connectedness (weak to strong). Systems cycle and change based on processes linked across scales that range from small and fast, to large and slow. As a system-level trait, resilience describes the structures and dynamics that may keep a system operating within a known range of values, or alternatively, the combination of circumstances that may lead to a system approaching and crossing a threshold and changing to a new state, with an associated set of emergent structures and feedbacks (Scheffer et al. 2012; Wilson et al. 2013). A system may subsequently transform, whereby a new system identity emerges based on a process of fundamental and significant change (Abel et al. 2006). These emergent conditions may represent tradeoffs for people and landscapes within an SES. Whether outcomes of transformation are negative or positive is due to the degree to which change is purposefully managed as opposed to experienced abruptly or unintentionally. Non-linear change or surprise events are potential triggers in the latter context whereby a tipping point or threshold is passed, causing significant transformations. Clark and Dickson (2003) suggest that there is greater potential for positive and sustainable outcomes when transformations are purposefully navigated. Significant Arctic research is ongoing currently to identify the social, institutional, governance and economic contexts under which this positive transformative capacity is facilitated (Berkes et al. 2003; Kofinas et al. 2013; Arctic Council 2016).

While resilience is invoked often as a framework to link social and ecological dynamics, the most detailed empirical operationalization of resilience concepts has occurred so far in ecological contexts (e.g. coral reefs (Hughes et al. 2003) and freshwater lakes, contexts that are non-Arctic (Carpenter and Cottingham 1997; Folke et al. 2004). Adger (2007) framed the concept of "social resilience" as distinct from ecological resilience. Resilient social systems were described in terms of their outcomes (i.e. sustainability), but also their ability to withstand or adapt to severe socio-economic and ecological disruptions based on sound institutions and governance structures (Adger 2000); Berkes and Folke (1998) carefully articulated the early challenges associated with linking social and ecological systems, but less research overall has systematically examined the resilience of integrated SESs. A significant challenge remains how to operationalize the capacity of any social–ecological system to remain functionally stable or robust under a range of exposures (Cumming et al. 2005). The very real challenges associated with analyzing and operationalizing resilience have sometimes led to assumptions that, if a system or a community is observed to have remained in place through time, it is de facto "resilient". These assumptions however ignore potentially significant cross-scale or temporal tradeoffs associated with change for households and communities across key indicators (for example, well-being, social cohesion, institutional capacity and flows of ecosystem services etc.) (Armitage et al. 2012; BurnSilver et al. 2017; Coulthard 2012); Olsson et al. (2004); Forbes et al. (2009) argue that if the goal is to understand the system traits (structure) that contribute to adaptation (system function) and resilience, as much effort should be placed on understanding key institutional and organizational landscapes as on ecological ones. These efforts translate into efforts that identify actions or frameworks that can "build resilience" into SES, for example adaptive co-management (Armitage et al. 2009; Armitage et al. 2011) or methodologies that create opportunities for co-learning and power sharing among diverse stakeholders and user groups (Kofinas 2005; Berkes 2009).

The Arctic Resilience Report (Arctic Council 2016) recently articulated a "state of resilience science" for the circumpolar Arctic. The framework as articulated is less focused on predicting who or what will be resilient in the context of ongoing processes of change in the Arctic, but rather how adaptive or transformational change may be facilitated based on more informed understandings of critical social–ecological dynamics occurring across multiple scales (Arctic Council 2013, 21). The questions "resilience of what, or whom, to what" remain critical in defining the social groups, institutional and ecological dynamics, interactions and scales at which questions of resilience should be posed and examined within SESs. Tanner et al. (2015, 23) suggest that consideration of livelihood resilience is a means to situate rights, justice, development and livelihood thresholds into conversations about transformation and sustainability more broadly. The broader recognition that resilience is a normative concept is now also percolating through resilience scholarship, giving rise to a new question, "resilience as defined by whom?" Issues of justice, equity and power are increasingly informing discussions regarding resilience transformations and visions for SES futures (Cote and Nightingale 2012; Fabinyi et al. 2014).

4.5 Arctic sustainability governance

In addition to the governance aspects in context of vulnerability and resilience as outlined above, a literature on Arctic sustainability governance is evolving, mostly since the establishment of the Arctic Council in 1996. Studies go also back further, starting with Mikhail Gorbachev's Murmansk speech in 1987 that triggered the Rovaniemi Process, which resulted in the Arctic Environmental Protection Strategy (AEPS) (Russell 1996), the predecessor of the Arctic Council. A focus has been on how Arctic cooperation has evolved over time and how the focus has evolved from environmental protection and conservation in the AEPS to the twin pillars of the Arctic Council that added sustainable development to the agenda, and which role the Council could play in the future in terms of fostering sustainable Arctic futures (Nord 2016; English 2013; Dodds et al. 2012; Axworthy et al 2012; Koivurova and VanderZwaag 2007; Koivurova 2010). Further studies have focused on the effectiveness of the Council in living up to its mandated tasks and in identifying new emerging issues (Young 2016; Kankaanpää and Young 2012).

Literature on Arctic sustainability governance can further be divided into more pan-Arctic governance issues, which mostly focus on marine governance (Berkman and Young 2009; Koivurova and Molenaar 2010; Tedsen et al. 2014) and region-specific governance issues, such as maritime delimitation and environmental management in the Barents Sea (Aven and Renn 2012; Henriksen and Ulfstein 2011). Again other literature focuses more on areas specific to sustainability governance; issues like fisheries (mostly in the area of conservation and management) (Molenaar and Corell 2009a; Molenaar 2009, 2012a, 2012b) and shipping (Molenaar and Corell 2009b; Molenaar 2011). A few studies investigated community-scale empowerment and "fate control" (Dahl et al. 2010).

Following the sea-ice minimum of 2007 and the increasing political interest in the Arctic, other theoretical frameworks have become more prominent in the literature, including attention to geopolitics and security (e.g. Keil 2014, 2015, Heininen and Nicol 2007; Baev 2007; Dittmer et al. 2011; Nilsson 2012; Huebert et al. 2012; Gjørv et al. 2013; Sorlin 2013) as well as to the role of media images in narrating the Arctic (Christensen et al. 2013). A promising new field of Arctic governance studies are examinations of the linkages between Arctic and non-Arctic processes and systems, concerning both social and ecological systems (Keil and Knecht 2017). Other relatively new concepts, such as "stewardship," have been forwarded as the most suitable ways of framing and addressing the political challenges (Griffiths et al. 2011; Arctic Governance Project 2010; Berkman 2010; Chapin et al. 2015). More recently there have also been efforts to link governance more explicitly to understanding the dynamic linkages between social and ecosystem aspects of SESs (Sommerkorn and Nilsson 2015).

5   Progress in Arctic sustainability research 2

Methodological advances

Accompanying progress in conceptual understandings of sustainability, methodologies employed in sustainability research have undergone considerable evolution in the last decade. Moreover, although Arctic sustainability research continues to rely on methods adopted from other regional contexts and follows the overall methodological trends in the field, the number and scope of studies that implement "Arctic-grown" methodologies is increasing. Arguably, we are at a point when Arctic sustainability research is offering and testing novel approaches and methodological frameworks, especially in respect to knowledge co-production, indicators building and community-based research. It is important to reflect on these achievements and contributions in order to determine existing trends, identify priorities for future research and define the role of Arctic sustainability research in global sustainability science of tomorrow. Below we discuss the following "epistemological transitions," which took place in Arctic sustainability research over the last 10 years: 1) transition to more integrated, inter- and transdisciplinary and mixed-method research approach; 2) transition from focus on sustainability as an outcome, to studying sustainability as a process; 3) transition to co-production of knowledge, including community-based approaches to research design, execution, assimilation and dissemination; and 4) development of indicators.

5.1 Transition to more integrated, inter- and transdisciplinary and mixed-method research

By definition, sustainability challenges represent complex questions that are broader than any one discipline can address. This statement now approaches a truism expressed in the literature. However, it implies a number of important points that have significant consequences for the conditions under which progress in sustainability science and practice may be achieved.

Because social and ecological systems are coupled, the delineation between social and ecological, or human vs. natural, is seen as increasingly arbitrary and obstructive to understanding the interactions, drivers, processes and thresholds of change characterizing SES through time and across space (Berkes and Folke 1998). Social–ecological system behavior is characterized by complex, non-linear dynamics, in which surprise and uncertainty are the norm, not the exception. These theoretical threads have combined in an epistemological call for science, research and collaboration to occur across disciplinary divides, i.e. to be at the least inter-disciplinary, and when possible, transdisciplinary. Segmented, sectoral or disciplinary thinking is limiting, as potential solutions that any single discipline can provide will be simplistic, narrow or lacking an appreciation of potential tradeoffs characterizing processes of change within linked social–ecological systems. As well, disciplines apply different epistemologies, frameworks and methods to ground their specific questions and concerns. Although discipline-based research reflects only a fraction of broader scale dynamics, these narrow findings may be applied widely to existing development or planning problems. Disciplinary tensions between the social and biophysical sciences are often alluded to, but equally challenging may be differences within the social sciences and humanities. There is a role for discipline-based science, to provide a great deal of depth in understanding of critical processes or mechanisms. However, fractured, disciplinary-based approaches will not be sufficient to address complex sustainability challenges in the Arctic.

There are key distinctions between interdisciplinary and transdisciplinary research epistemologies. Interdisciplinary research involves the "integration of a number of disciplines into a coherent research cluster providing a new framework for understanding," whereas "transdisciplinarity" may entail "the inclusion of other forms of knowledge than scientific knowledge in the research process" (Holm et al. 2013, 29). Klein (2004) highlights the difference between interdisciplinary "synthesis" – implying unity – as opposed to "coherence" between multiple perspectives that continue to exist (521, and see Ramadier 2004). Transdisciplinarity implies a process of seeking out diverse perspectives, then deliberating and communicating across these boundaries (Ramadier 2004). Transdisciplinary science is therefore an emergent outcome of collaborative process, one that ideally co-produces new knowledge and understanding of what sustainability means and how it could be achieved. Collaboration here is conceptualized broadly as a methodological approach, one that implies "making sense together" across diverse stakeholder groups and academic disciplines (i.e. policy makers, communities, business entities, and scientists) (Klein 2004, 521). Conflict and disagreement are critical to this process as initial assumptions are deconstructed and debated across stakeholder groups and disciplines that have widely varying assumptions and goals, and importantly, very unequal levels of decision-making power. This conceptualization of transdisciplinarity has broad implications for understanding sustainability. It positions sustainability as both a process of articulation across differences, and a normative outcome.

5.2 Conceptualizing sustainability as both process and outcome

In recent years there has been a shift from a singular focus on sustainability as a normative outcome, to understanding sustainability as a process. The process has to be examined from the perspective of desired outcome(s), but also (and sometimes more importantly) from the perspective of identifying mechanisms, contexts and actions that contribute to new forms of perspective-taking, innovation and collaboration convened around intractable sustainability challenges (Westley et al. 2011). A methodological focus on key characteristics of process can itself lead to co-learning around an issue, innovating solutions or developing capacities to do so in the future. These processes require careful design and facilitation, as eliciting and integrating diverse sets of knowledge and values implicit in different framings of "sustainability problems" is a significant challenge (Leach et al. 2010). Implicit in this effort is an engagement philosophy that recognizes the deep uncertainty characterizing all decision-making processes (Wise et al. 2014), but engages with diverse stakeholders to design pathways inclusive of broader narratives on values, tradeoffs, solutions and sustainability (Haasnoot et al. 2013). Interest in the process of sustainable development is also associated with a rise of "action research" that is designed to deal with immediate circumstances rather than with possible distant outcomes. Partially, this trend is determined by rapid changes in Arctic societies and environments, and the necessity to quickly understand and address specific problems in sustainable ways. Examples of this approach include community climate-change action plans (Ford et al. 2010; Ford and Goldhar 2012; Wolf et al. 2013), research targeting community-defined issues (e.g. housing, food security, water, transportation, etc.) (Ridlesperger 2014; White at al. 2007). In most cases process-oriented research is community-driven, but this trend is also becoming apparent in regional, circumpolar and national assessment processes (such as Adaptation Actions for a Changing Arctic).

An interesting element of this "process-based" research agenda is a greater appreciation for the role of conflict, success and failure in achieving sustainable development outcomes (Leach et al. 2013). A careful analysis of success stories and failed projects can be very informative in respect to improving our knowledge about the process of sustainable development, its agents and factors. For example, this may involve an examination of institutional context, socio-economic dynamics and personal characteristics of individuals involved in the process. The critical role of these factors in the arctic context are yet to be explored.

5.3 Co-production of knowledge and community-based research

Knowledge co-production has been identified as a prerequisite for sustainability transformation (Pohl et al. 2010). The term co-production refers to a joint process between scholars and various partners (communities, governments, private industry) of planning, carrying out, and disseminating research (also see Trencher et al. 2013). Dimensions of co-production include the gathering, sharing, integration, interpretation, and application of knowledge (Armitage et al. 2011). Co-production can take place in a community setting, though this must not necessarily be the case. While the forming of partnerships between northern communities and academics is not a new phenomenon, there are recent trends towards more collaborative inter-, and transdisciplinary research that aim to be truly reflective of Northern people's priorities (Tondu et al. 2014). In part this is due to recognizing the importance of social dimensions to various issues in the Arctic (Adams et al. 2014). When pursued with indigenous partners, such co-production of knowledge can also play an important role in a process of decolonizing research, although the process is rife with challenges of power relations and hierarchical acceptance of multiples types of knowledge (e.g. Watson and Huntington 2014; Wråkberg and Granqvist 2014).

Knowledge and research that are community-informed are indeed better suited to address complex sustainability challenges. In addition, there are important ethical implications for research occurring in Northern indigenous spaces and places that can in part be addressed through processes of co-production (Kershaw et al. 2014). Due to increasing autonomy of indigenous peoples, co-production of knowledge may not only be encouraged, but a requirement as part of the protocol for research licensing in Arctic regions (ITK/NRI 2007; Pearce et al. 2009). While there has been an increasing awareness of the importance of knowledge co-production in the Arctic, there are still important challenges to be overcome. Desbiens (2010) notes, for example, that climate-change research in the North predominantly focuses on male knowledge and processes ("big game trumps berries"). This leaves the knowledge and teaching of indigenous women largely undocumented, while at the same time imposing preconceived, and possibly ill-suited and inappropriate, research frameworks and questions on communities and individuals. Watson and Huntington (2014) observe that too often collaborative research that includes TEK/IK still only treats such knowledge as local, failing to consider the epistemological and ontological levels of IK – and claims that this approach is "vulnerable to committing the same erasure of indigenous peoples that occurred across colonial science." (Watson and Huntington 2014, 733). Wråkberg and Granqvist (2014) reflect that differences in how traditional and scientific knowledge are generated, enacted, and stored challenge the equitable co-production of knowledge. Armitage et al. (2011), among others, identify further challenges of knowledge co-production including: the role of power, shared understanding, and normative context. Ensuring the equality of partners, and respecting indigenous knowledge and methodologies alongside "Western science" knowledge and methodologies remains challenging to many partnerships (e.g. Hart 2010; Mutua and Swaderner 2004; Rigney 1999; and Rasmus 2014 for Alaska specifically). Indigenous requirements for ethical research have begun to be articulated in ways accessible to outsiders (e.g. AFN 2009; ITK/NRI 2007); they sometimes diverge from requirements stipulated by professional/institutional codes of conduct (e.g. those of Research Ethics Boards). Sustainability of research in the Arctic requires attention to both the equity of processes and the ethics of knowledge co-production.

Co-production of knowledge is also relevant in relation to a range of other actors, including decision makers in municipalities and regional government as well as private sector actors. These are less developed as explicit co-production endeavors. However, there is a base to build on from local studies of adaptation and adaptive capacity that are often carried out in ways that include engagement with local actors. Ventures into co-production that involve commercial actors come with somewhat different ethical issues than community engagement that relate to power relationships. Assessment processes, such as the assessment carried out under the auspices of the Arctic Council can also be seen as processes for co-production of knowledge. Ownership of results (and process), including how results are made public, are issues that warrant further discussion when moving into this mode of research.

6 Progress in Arctic sustainability research 3

Sustainability indicators

Indicators are metrics used to determine the current state and, when monitored over time, the trajectory of change for a given domain. Indicators are often used to gauge progress toward specified goals (so-called performance indicators). Sustainability indicators allow tracking of trends toward sustainability goals, and identification of the need for intervention in cases where the trajectory of change is in an undesirable direction. Such indicators can be useful in supporting both the design and implementation of policy. The absence of measures to monitor progress toward sustainable development is seen as a major impediment to its achievement (Bossel 1999).

To be useful, indicators must be measurable (quantitative indicators are usually preferred), based on data that are not too expensive to collect, sensitive (responsive to change) and understandable. Indicators based on data that are readily available are often preferred (Larsen et al 2010). It is challenging, however, to develop a limited set of indicators that adequately address the multifaceted nature of sustainable development, incorporating environmental, economic and social aspects (Hamilton and Lammers 2011). While indices, or compound indicators, are often seen as a way to address such complexity, such indices themselves pose challenges in terms of what the relative weightings of the composite indicators should be, and to what extent the indicators are inter-correlated. As of yet, there is no Arctic indicator set or index that includes social, economic and environmental aspects, to offer a tool for tracking sustainable development in the Arctic (Ozkan and Schott 2013).

6.1 Global sustainability indicator initiatives

The Bruntland Report's (1987) call for sustainable development stimulated the need for ways to measure whether progress is being made in moving toward this goal. Five years later, Agenda 21, emanating from the UN Conference on Environment and Development in 1992 (aka the

"Earth Summit") called for national governments to develop National Sustainable Development Strategies (NSDSs), with indicators as a feature of these to track progress. Following this, the UN Commission on Sustainable Development approved a Work Programme on Indicators of Sustainable Development. Many countries adopted the indicators developed under this aegis as the basis on for their own sustainable development indicators. Further, in 2000, the UN Millennium Declaration called for establishing Millennium Development Goals (MDGs) – and indicators to track progress toward their achievement. These and other calls have given rise to the development of numerous indicators and indices related to sustainable development, such as the Environmental Sustainability Index, the Environmental Performance Index, Genuine Progress Indicator/Index, Ecological Footprint calculations and Living Planet Index. The challenge has been to develop a small set of indicators that meet the criteria discussed above, and, as noted, to incorporate social, economic and environmental measures. The adoption of the global Sustainable Development Goals in 2015[1] is giving rise to further efforts in this field.

While several Arctic countries have developed NSDSs, along with indicators to support these, these are national-level indicators, not specifically designed for the Arctic. The first Arctic Human Development Report (Nilsson et al. 2004) emphasized both the need to track human development over time, and the need for a limited set of indicators that reflect key aspects of Arctic human development. Specifically, the report identified key aspects of well-being that Arctic residents greatly value: fate control, closeness to nature and cultural vitality (Nilsson et al. 2004, 240). While the first AHDR called for indicators of Arctic human development (not sustainable development), its recognition of the need for a distinctive regional approach to such indicators suggests that indicators developed for NSDSs may not be wholly relevant or adequate to the Arctic regions of these states.

As noted in Chapter 3, ICARP II Science Plans 1 and 10 called for the development of measures or indicators for sustainability, sustainable development and resilience. Initiatives to monitor environmental change in the Arctic via indicators have burgeoned over the past decade. The Arctic Monitoring and Assessment Programme (AMAP) has long monitored contaminant level in the Arctic and used this data as a basis for assessment of pollution trends. The Circumpolar Biodiversity Monitoring Program (CBMP), developed by the Arctic Council's Conservation of Arctic Fauna and Flora Working Group (CAFF), responded to the Arctic Climate Impact Assessment's (ACIA) call for improving and expanding biodiversity monitoring in the Arctic (ACIA 2005). CAFF recently *published Arctic Biodiversity Trends 2010: Selected Indicators of Change* (CAFF 2010), a suite of indicators developed by the CBMP, as part of its Arctic Biodiversity

Assessment initiative. CAFF is also working with the Norwegian Institute for Nature Research to develop an Arctic Nature Index.[2]

On the social and economic side, we also can identify several projects. Most notably, the Arctic Social Indicators project (Larsen et al. 2010; 2015), is a direct response to the first AHDR's call for a system for tracking trends in human development in the Arctic over time, through the identification of a set of indicators (AHDR 2004,11). The ASI project developed a set of indicators for the domains of Fate Control, Closeness to Nature, Cultural Vitality, Material Well-Being, Health and Well-Being, and Education (Larsen et al. 2010). The goal was for the indicators to reflect key aspects of human well-being in the Arctic, including those aspects specific to the region, as identified in the first AHDR (2004). During a second phase of the ASI project, the proposed indicators were tested through application in several regions of the North (Larsen et al. 2015). While an important start, the ASI indicators need refinement. Moreover, as the second AHDR concludes, we need indicators that "better reflect the impact of global changes, including the complex interactions between biophysical changes and changes in human systems" (Larsen and Fondahl 2014; see also Ozkan and Schott 2013).

An Arctic Observing Network Social Indicators Project (AON-SIP) was also initiated under the aegis of the SEARCH (Study of Environmental Arctic Change) project (Kruse et al. 2011). This project was less focused on generating indicators, but rather set out to compile existing data relevant to specific focus areas (fisheries, subsistence, tourism, and resource development) and to make recommendations on how to improve the database (Kruse et al. 2011, 4; see also related articles in *Polar Geography* 34 (1–2)). Significant overlap between ASI and AON-SIP members (including both AON-SIP principal investigators being on the ASI team) enhanced complementarity of the two projects.

Numerous regional-level initiatives for measuring human development have also resulted in proposed indicators. The Social Cohesion Index (Duhaime et al. 2004), for the Canadian Arctic, measures several facets of human development: social capital, demographic stability, social inclusion, economic inclusion; and both community and individual quality of life. The Centre for Strategic Research of the Sakha Republic (Yakutia) has initiated a project, "Humans in the Arctic," to establish an indicator and monitoring framework for the Russian North, which could eventually be expanded to the Circumpolar North, and is currently examining 52 indicators, including those suggested by the ASI project, for feasibility (personal communication, 19 February 2015). A social and economic indicators project for the Inuvialuit region in northwest Canada was established under the aegis of the large-scale Resources and Sustainable Development in the Arctic (ReSDA)

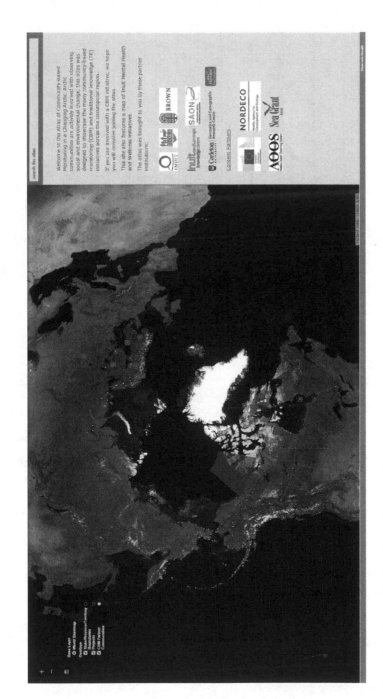

Figure 6.1 Community based monitoring projects

Source: http://www.arcticcbm.org/index.htm

research project.[3] Novikova and Stepanov proposed a suite of 43 indicators to measure quality of life for Sakhalin Oblast, partly based on data collected by the government, partly requiring surveys and interviews of the population (Novikova and Stepanov 2010).[4]

Local efforts to develop indicators for environmental monitoring in the Arctic have also increased with the proliferation of community-based monitoring projects, and are too numerous to mention here. However, we do cite the Atlas of Community-Based Monitoring and Traditional Knowledge in a Changing Arctic (Figure 6.1). Also noteworthy is ELOKA (Exchange for Local Observations and Knowledge in the Arctic), an initiative to facilitate the exchange of data obtained from monitoring. Its focus is offering support in data management and exchange to community-based monitoring systems. CACCON (Circumpolar Coastal Communities Observatory Network) was recently established to facilitate and interconnect observations in coastal social–ecological systems in the Arctic. To date much of the information collected at the local level, including the traditional knowledge documented, has been environmental in nature.

Perhaps the largest regional-scale attempt at producing Arctic sustainability indicators was led by Votrin, who developed a suite of 30 sustainability indicators, including social (9), environmental (14) and economic (7) measures, for the Russian Arctic (Votrin 2006). Three indicators are Arctic-specific (all of them environmental). Unfortunately Votrin's work remains unpublished. Moreover, while the Russian Arctic constitutes a large part of the whole region, and while many of Votrin's indicators might serve well for the rest of the Arctic, we remain without an integrated set of indicators specific to the Arctic region as a whole, to be used for tracking sustainable development.

6.2 Challenges to developing Arctic sustainable development indicators

Challenges to developing sustainable development indicators for the Arctic include data availability and quality issues, but also issues of common definitions for indicators across the Arctic, in order to ensure comparability, and consistent data collection, both across space and time (cf. Pintér et al. 2005). While data collection and quality in Arctic states is more advanced than in many other regions, there are still serious lacunae (and very concerning declines in some data collection, such as Canada eliminating its mandatory long-form census in 2010). Moreover, the difficulty of linking "fundamentally disparate data from social sciences and physically-based gridded data fields" remains a challenge, although one that has begun to be addressed (Hamilton and Lammers 2011, 121).

Notes

1 https://sustainabledevelopment.un.org/
2 http://www.caff.is/indices-and-indicators/arctic-nature-index
3 http://yukonresearch.yukoncollege.yk.ca/resda/projects/inuvialuit-indicators-project/
4 While Novikova and Stepanov cite the importance of considering the correlation of the ecological situation with well-being (Novikova and Stepanov 2010,38), the proposed indicators do not include environmental ones.

7 Different spatial scales, global, national, regional, local, and their interconnections with Arctic and non-Arctic regions

The ongoing transformations occurring in Arctic regions are deeply intertwined with regional and global processes, since the Arctic is both affecting and being affected by broader regional and global processes reaching beyond its southern borders. Various natural sciences disciplines have produced a vast amount of literature on the feedback loops between the climatic, environmental and atmospheric systems of the Arctic and the rest of the globe (IPCC 2014). Among others, addressed topics include global–Arctic links concerning climate change, contaminants, drivers and effects of sea-ice melting and possible effects on weather and ocean circulation, and the possibility of increased emissions released from permafrost thawing (e.g., ACIA 2005; Chapin et al. 2012; McGuire et al. 2006; Vihma 2014).

7.1 Multi-scale sustainability studies within social science

Interdependencies between social science aspects – such as economic developments, legal frameworks and patterns of livelihoods and general development of societies – both on different spatial scales within Arctic regions and between Arctic and non-Arctic regions have so far received less attention, especially in regard to sustainable development. Among the few exceptions is Young's 1992 chapter on "Sustainable Development in the Arctic: The International Dimensions" (Young 1992a). Further, while studies on multi-scale interaction within certain Arctic regions and on certain issues like climate change and environmental policy exist (e.g. federal-subnational interactions in Canada and the US) (see Nilsson 2009b; Byrnea, Hughesa, Wilson, and Kurdgelashvili 2007; May, Jones, Beem, Neff-Sharum, and Poague 2005; Nadasdy 2003; Rabe 2004), there is a need for bringing findings of such regional studies together for inter-regional multi-scale comparisons.

There is a stronger understanding of the history of Arctic/non-Arctic connections, highlighting different Arctic regions' early embeddedness in economic flows – such as in the early whale hunting and fur trade business, fisheries, and exploration and exploitation of mineral resources (e.g. Grant 2010; Heininen et al. 2010; Sale and Potapov 2010, Vaughan 2007; Stuhl 2016) – and studies on the Arctic as a strategic theatre during World War I and II and the Cold War (see Heininen 2010; Heininen et al. 2010; Young 1985, 1992b).

Beyond history, scholarly work on Arctic international relations and Arctic governance has focused on linking Arctic and international scales. But these research areas have long focused on the role of the state and region-wide inter-governmental politics among the five Arctic coastal states (Canada, the US, Russia, Norway and Denmark/Greenland) when addressing the issue of Arctic sustainable development (Byers, 2009; Hough 2013; Murray and Nuttall 2014; Stephens and VanderZwaag 2014; Stokke and Hønneland 2006; Tamnes and Offerdal 2014). These studies often implicitly adopt a regional lens emphasizing circumpolar solutions to circumpolar problems and leaving aside both formal and more informal steering mechanisms on other scales below and beyond the state-level and outside the Arctic region (Berkman and Vylegzhanin 2013; Tedsen et al. 2014; Keil and Knecht 2017).

While some recent trends in Arctic International Relations research reach out to non-Arctic actors – especially the role of Asian countries in and for the Arctic as well as international actors like the European Union (e.g. Airoldi 2008; Chen 2012; Jakobson and Peng 2012; Jakobson 2010; Koivurova et al. 2011; Chaturvedi 2012; Manicom and Lackenbauer 2013; Neumann 2010; Røseth 2014) – these remain predominantly studies on the state-level and with little explicit reference to the role and influence of these actors on Arctic sustainable development. One notable research project on the role of non-Arctic actors (but little on their role for Arctic sustainable development specifically) is the 2012–2015 AsiArctic program (www.asiarctic.no) run by the Fridtjof Nansen Institute and the Norwegian Institute for Defence Studies in cooperation with partners from India, South Korea, Japan, and China. Like the majority of research efforts, the focus is on political, strategic/security and interest related aspects of the role of non-Arctic actors in the Arctic (especially in relation to the opportunities associated with the increasing accessibility of the Arctic). One notable exception is the study by Cavalieri et al. (2010), analyzing the European Union's environmental footprint on the Arctic.

Generally, the state-centered view of many approaches addressing global relevance of Arctic developments has often emphasized the importance of the Arctic to traditional understandings of security and the exercise of state sovereignty instead of aiming at a more nuanced view on the sustainable management of Arctic space and global commons also on local levels (see Kraska 2011; Tamnes and Offerdal 2014; Zellen 2013). As important as this generation of studies continues to be for the understanding of international governance in the circumpolar North, it disregards important parameters of the Arctic's political geography and contemporary developments in the age of globalization. Most importantly, Arctic change is not a sole regional challenge. The Arctic is a "globally embedded space" (Keil and Knecht 2017) and cannot be detached from transnational and international patterns and processes that interfere with the region and co-determine the outlooks for Arctic sustainable development (Berkman and Young 2009).

7.2 Multi-scale sustainability studies involving natural and social science

In addition to the lack of inter-scale studies with a specific focus on sustainability within social sciences, there is a lack of research on multiple scale issue areas involving natural and social processes, as well as humanities. Health issues are one example where cross-disciplinary efforts are high in demand (e.g. pollutants and their long-range travel, social and political practices related to production of these pollutants, and governance responses thereto). This means tapping into the challenging area of strong or "radical" interdisciplinary research, i.e. across the realms of natural and social sciences and humanities. While interdisciplinary research within natural or social sciences (what Holm et al. (2013) referred to as "contained interdisciplinarity;" e.g. between biology, chemistry, and oceanography or between history, political science, and anthropology) does already exist in some research efforts (for example, within the "Transformations to Sustainability" Program of the International Social Science Council), the strong or "radical" (ibid.) variant across natural and social science borders is still in its infancy. Further, while some research efforts with strong interdisciplinary approaches are discernible (such as the in the 2011–2015 Arctic Climate Change, Economy and Society (ACCESS) project) they often remain rather multidisciplinary, i.e. with disciplines separated in different working groups with little effort to integrate between the disciplines. Such projects produce highly important and relevant output, of course! But sharing a common, concrete objective and working towards a common outcome of a research effort across disciplines has so far rather been the exception among interdisciplinary projects.

7.3 Avenues for future research at different scales

Against the background of the general gaps as outlined above, the following highlights a number of research avenues for studying Arctic sustainable development on different spatial scales and in relation to interconnections within the Arctic and with non-Arctic regions.

First of all, we need to better understand and address the myth(s) of Arctic exceptionalism, which are widespread in Arctic studies. While in many respects the Arctic is surely unique and differs from other regions, there is a lot to learn and gain from a better understanding of lessons learned from Arctic regions, which could be beneficial for other regions outside the Arctic. Similarly we can learn and understand more about the Arctic when studying other places and regions. Methodologically, there is thus a need for more comparative efforts with non-Arctic case studies.

Research efforts could increasingly focus on understudied issue areas with global–national–regional–local linkages, in order to better understand outlooks and pathways for Arctic sustainable development as well as the Arctic's role in global processes and sustainable development challenges. Examples for such issue areas include Arctic resources (oil and gas, mining, shipping, fisheries), tourism, international migration, and the Arctic's role in the global food-soil-water-energy nexus and related challenges such as water shortages, and energy security.

For example, in relation to Arctic resources, more extensive studies are needed to include the various scales and stakeholders involved in Arctic resource development and the effect this may have on sustainable development in the region and beyond. Studies like these would also provide a better understanding of the Arctic in global economic pathways like production, consumption and trade flows of various commodities.

As another example, there is a need for a better understanding of global geopolitical developments affecting the Arctic and how they play out at different scales. Related to this, we need deeper insights into the different use and utilization of security and sovereignty concerns on different scales and what this means for sustainability understandings and prospects for sustainable development. In this and other realms, there are thus far largely untapped synergies for better inter-scale understandings, such as between anthropologists with extensive expertise on the local level and political scientists usually more engaged in processes on national and international levels. Such studies would also provide greater insight in the dynamics between different scales as well as the power relations between them.

A better understanding of the dynamics of different scales could also be achieved through systematic analyses of the coverage and standing of the Arctic in different media sources (see Christensen 2013 as an example). Research on the Arctic in the context of globalized media, the development of attention levels towards the Arctic, and thus the connectivity of the Arctic with the globe and the travel of ideas and images between different scales promise valuable input, not least on the level and extent of global cultural connectivity and possible linkages between regional-specific and global agendas for sustainable development. A new area of research in this regard is the development of communication in and about the Arctic, including the global communication about the Arctic in next to real-time social and online media networks, and possible effects on social capital (vertical in additional to horizontal; not only bonding but also bridging). Contributions from cultural studies and other humanities fields will be critical here.

Pursuing the above outlined avenues for research on Arctic sustainable development on different spatial scales and in relation to interconnections within the Arctic and with non-Arctic regions will also affect methodologies for Arctic research. The rather understudied links between multiple scales within the Arctic and beyond is also reflected in limited stakeholder engagement in Arctic research (see Chapter 5). Many studies – if engaging stakeholders in a truly interdisciplinary way at all – are often limited to stake- and rights-holders from within the Arctic, and seldom reach out to Arctic stakeholders residing outside the Arctic. So there is a lot to engage in concerning the establishment of new and strengthening of existing Arctic stakeholder connections, both from within and outside the Arctic region.

8 Agenda 2025

Perspectives on gaps and future research priorities in Arctic sustainability research

8.1 Key developments and progress in Arctic sustainability research

A critical overview of directions and achievements in the last 10–15 years of sustainability research undertaken in this book identified areas of substantial progress and considerable gaps in emerging Arctic sustainability science. In general Arctic sustainability research demonstrates growing theoretical and methodological strengths. On the academic side sustainability and sustainable development studies from the Arctic are now not only "catching up" with similar research outside the region, but in many ways offer pioneering approaches to theory, methodology and implementation of sustainable development. Arctic-based research provided substantive and substantial input to our understanding of coupled human-environment interactions from multiple perspectives, especially from standpoints of resilience, adaptation and ecosystem stewardship approaches. Arctic researchers made contributions in developing social and environmental indicators and implementing community-based methodologies of community-oriented, community-relevant research. It is our assessment that Arctic sustainability research is becoming a leading regional contributor to global knowledge about sustainability.

Progress in theory and emerging features of Arctic sustainability science

Although studies in sustainable development and sustainability in the Arctic showed rapid development in many different directions, the following are the key elements we consider most notable. First, the theoretical crystallization of Arctic sustainability research has been tied to the evolution from generic sustainability science approaches or purely localism to the notion of Arctic uniqueness among global connectedness. The general path of theoretical evolution was also in respect to systematic understanding and

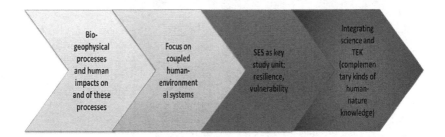

Figure 8.1 Conceptual shifts in Arctic sustainability research (2000–present)

interpretation of biogeophysical and human processes in their interaction. The progress in the last decade was quite rapid and includes several major conceptual shifts (Figure 8.1). The first and earliest shift was from "loose coupling" that considered primarily "human dimensions" of environmental change and impacts of humans on biogeophysical processes to the focus on coupled human–environmental systems. The next shift has firmly placed social–ecological systems at the center of the inquiry and adopted resilience and vulnerability frameworks to their analysis. The third shift, which is still happening now, is instigated by incorporating traditional knowledge in sustainability research and emerging complementary and integrated systems of human–nature knowledge. On the other hand, the scope and strengths of sustainability research in the Arctic may point to the emerging new 'stream' of inquiry, the Arctic sustainability science.

What are the core characteristics of the contemporary sustainability and sustainable development research in the Arctic? In other words, what are the features of the emerging Arctic sustainability science? These points are outlined below.

Emergent features of Arctic sustainability science

- Addresses the long term interactions of people and ecosystems

  o Social–ecological Systems (SES): primary unit of analysis
  o Inter- or transdisciplinary
  o But challenges implied in systematically addressing these linkages

- A complex systems focus: "wicked problems"

  o Non-linear changes and transformations
  o Uncertainty in drivers and outcomes
  o Solutions often generate or inform other problems

- Problem-focused: addresses "grand challenges"
    - Climate change: the Arctic as a "bell weather" region
    - Well-being and economic development
    - Integrating multiple ways of knowing about systems
    - Adaptive Co-management, adaptive governance

First, it is the focus on social–ecological system as the primary unit of analysis, a concern with long-term interactions of human and natural systems. Consequently, this leads to a deep engagement of inter- and transdisciplinary approaches in research and fundamental interest in systematically addressing linkages in SES. Secondly, Arctic sustainability research deals with "wicked problems" associated with non-linear processes, changes and transformations, as well as with uncertainty in drivers and outcomes. Thirdly, Arctic sustainability research is largely problem-focused and orients itself as addressing 'grand challenges', such as climate change, well-being and economic development, integrating Western and traditional knowledge systems to aid decision-making, adaptive co-management and governance.

Progress in methodologies and four epistemological transitions

In respect to methodological advances, we can point out three key conclusions. One is that Arctic sustainability research actively employs methodologies that stem from various fields, thus resulting in rigorous mixed methods, inter- and transdisciplinary methodological apparatus. Another is the increased evidence of novel cutting-edge approaches and methodologies coming from the Arctic. Arctic sustainability research is offering and testing novel approaches and methodological frameworks. Examples of these include knowledge co-production, indicators development, community-based research methods and so on.

We also identified four epistemological transitions that have happened or are ongoing and reflect the direction(s) in which methodologies are evolving. Transition one constitutes the move to integrated trans/interdisciplinary and mixed methods research. This transition is determined by theoretical evolution (Figure 8.1) to the study of coupled systems (SES) using methods from multiple discipliners, which require integration. Integrated and mixed methods approaches appear to be most effective in the literature we analyzed. Transdisciplinary scholarship goes beyond interdisciplinarity in involving and engaging a variety of stakeholders. Rather than having individual disciplines contribute separately, it tries to weave in varying methodologies to truly get at the nature of complex, or wicked, problems.

Transition two is the re-orientation towards looking at sustainable development as an outcome to studying it as a process. A growing understanding is that sustainable development is concerned with solving problems that are sometimes not fully understood, or even identified. In addition, the interest to the process of sustainable development is also associated with a rise of 'action research' that is designed to deal with immediate circumstances rather than with possible distant outcomes. This trend is partially determined by rapid changes in Arctic societies and environments, and the necessity to quickly understand and address specific problems. An interesting element of this "process-based" research agenda is the understanding of success and failure in achieving sustainable development outcomes. A careful analysis of success stories and failed projects can be very informative in respect to improving our knowledge about the process of sustainable development, its agents and factors. For example, this may involve an examination of institutional context, socio-economic dynamics and personal characteristics of individuals involved in the process.

Transition three is the rise of knowledge co-production as a central epistemological paradigm. Knowledge co-production has been identified as a prerequisite for sustainability transformation. "Co-production" refers to a joint process between academics and various partners (communities, governments, private industry) of planning, carrying out, and disseminating research. Dimensions of co-production include the gathering, sharing, integration, interpretation, and application of knowledge. Knowledge and research that are community-informed are indeed better suited to address complex sustainability challenges. In addition, there are important ethical implications for research occurring in northern indigenous spaces and places that can in part be addressed through processes of co-production.

Transition four is the evolution of indicators research from its initial concern with either social or environmental indicators towards integrated, joint systems of indicators. The 'rise of indicators' research is partially related to the connection of sustainability scholarship with vital stakeholder interests and decision-making in the face of "grand challenges" and immediate action needs. The systems of indicators improved considerably on both social and physical sides, and are becoming components of the integrated apparatus that is designed to measure and characterize dynamic and coupled social–ecological systems.

8.2 Key knowledge gaps

Previous sections of this book discussed most important achievements and assessed progress in Arctic sustainability research but also remaining gaps in Arctic sustainability research. From these overviews we have identified a number of key knowledge gaps as suggestions for future pathways for Arctic sustainability research. Some of them are long standing and have already been articulated in ICARP II Science Plans. Others are products of evolving disciplines, changing methodological frameworks and our improved understanding of the nature of Arctic social–ecological systems. In the concluding section we aim to describe a number of key elements that will be important for future research planning. They include knowledge gaps, both general and Arctic specific, research priorities for the next decade and actionable research items that specific directions researchers and funding agencies could pursue to serve these priorities.

General gaps (relevant within and beyond the Arctic)

General knowledge gaps are the unresolved deficiencies in our knowledge about sustainability and sustainable development in general, irrespective of regional application. In other words these are non-Arctic specific gaps that would require the efforts of all sustainability scholars, including Arctic sustainability researchers, to be filled. The following general knowledge gaps were identified as they have been reflected within the Arctic sustainability scholarship:

- **Insufficient research dealing with historical understanding of sustainable development in the Arctic**. We were unable to find literature that would be covering this topic exhaustively. There are multiple histories here, including history of science, policy and traditional societies. The importance of history is directly related to the legitimacy of the sustainable development concept in the Arctic, and only deep understanding of its beginnings and implementation may ensure the decolonizing nature of sustainable development. Historical awareness alongside other characteristics of Arctic sustainability research (such as community-based nature, knowledge co-production paradigm, etc.) will be the primary ways to avoid what one Arctic resident summed up by saying: "They first wanted to modernize us, now they want to make us sustainable."
- **Insufficient examination of different spatial and temporal scales, limited inter- and intra-scale linkages**: Palsson et al. (2013, 7) note how "acts that are legal, and sometimes even virtuous, under the current economic system, because of their contribution to job-creation and economic growth, have flip sides that, if they were registered as instantaneous acts, would be criminal"[1]. Likewise, acts that contribute to sustainability at the community level may not be at the regional or international level, as acts that appear sustainable over generations may seem to compromise immediate accomplishment of sustainability.

- **Weak linkages between spatial and temporal studies.** Most studies deal with either space or time and do not explore space-time linkages to the extent necessary to describe the dynamic and scale-dependent nature of SES and the sustainable develop processes.
- **Limited although improving integration of methods and disciplines,** including interdisciplinarity and, co-production of knowledge. Although we observed the epistemological transition towards such integration, still much remains unknown and untapped in respect to mechanisms and modalities of inter- and transdisciplinary research.
- **Relinking social–ecological systems to the top pillars of sustainable development.** Although there has been a change in understanding of sustainable development through the prism of socio-ecological systems, it should be re-linked to the three pillars of sustainable development (economics, ecology, equity) with a renewed re-emphasis on interdependence rather than competition between them (a "holistic approach").
- **Emerging but yet underdeveloped notions of the new study triangle**: economics, ecology, and equity. The stylized understanding of sustainability has been expanded by incorporating various other elements, such as ethics, into the scope of consideration. We need to look more deeply into the foundations of the three traditional pillars of sustainability, as well as the interdependence among them. In addition, it is argued that "trade-offs" between the "pillars" of sustainability may not be as important as the integration that focuses on interdependence rather than competition between them to achieve sustainability outcomes.
- **Lacking connectivity between social and physical indicators for sustainability.** Although we experienced a boom of indicators research in the last decade, there is still an acute shortage in frameworks that successfully integrate qualitative and quantitative indicators of social and biogeochemical components of SES.
- **Failure to address inter-generational and gender issues**: To date, we have a limited understanding of the gendered nature of vulnerability to climate and other change in the Arctic, and an inadequate understanding of the gendered dimensions of adaptation and resilience. We need to know more about gendered differences in cognition and behavior related to rapid socio-ecological change. Differential migration, educational attainment, and capacity for adaptation across genders influence the sustainability of households and communities in the Arctic.
- **Limited connectivity between conceptual work and empirical work.** Current scholarship is not fully utilizing its capacity to generate and guide action as it could. Despite advancements in action research in the last decade, the researchers still need to look more closely at transformative change in terms of output, outcomes and processes.

- **Minimal role of humanities in sustainability research**. In interdisciplinary efforts, social sciences and humanities research are often relegated to "an auxiliary, advisory and essentially non-scientific status" (Holm et al. 2013, 26). Indeed, humanities research is often fully absent from such collaboration. Yet humanities are sorely needed to address sustainability transitions. Contributions from communications, cultural studies, ethics, history, law, literature, linguistics, and philosophy are needed to fully understand how to identify, promote, incentivize and reward sustainable behaviors and the barriers to them. We know that people's attitudes and actions are far less affected by rational arguments than by emotional, instinctive reactions. We need to understand the cultural and cognitive factors, the personal and societal motivations, as well as the institutional conditions that contribute to, or thwart sustainable development. As noted above, the very concepts of sustainability and sustainable development undergo metamorphosis depending on the translation of key terms. Linguistic research into understandings and nuances of these concepts and challenges of communication is critical. Research into semantic clusters and metaphors reveal how people, both within and beyond the Arctic think about the Arctic and its future. In addition, indigenous languages often convey epistemologies and approaches that are absent or poorly expressed in colonizing languages.

We also identified research gaps that pertain to issues unique or specifically important in the Arctic. These gaps will have to be filled by the efforts of the Arctic scholarship and they also constitute a considerable part of our recommendations on how to move forward.

Arctic-specific gaps:

- **Lacking knowledge about urban areas, urban–rural connections and dynamics.** Arctic sustainability literature provides very limited focus on urban areas and urban–rural relationships. Although there have been a number of studies, they are scattered and sometimes devoid of a clear sustainability perspective, or are focused on sustainability-related concerns – such as climate change policy – but don't specifically address sustainability. Most case studies deal with selected rural settings without connecting to other rural contexts and linkages to urban realms. In the meantime, the majority of Arctic residents are urban dwellers.
- **Limited knowledge about non-indigenous people in the Arctic.** In the same vein the vast majority of sustainability literature is dealing with indigenous communities and societies. While this emphasis is understandable

and appropriate, and it must continue, more knowledge could be developed about non-indigenous societies. There is not enough engagement with scholarship outside the Arctic in respect to these societies.

- **Extremely rudimentary understanding of non-resource, non-traditional economies (such as knowledge economy, arts and crafts, etc.).** Most research is focused on sustainable development in the context of either resource (renewable or not) or traditional economy, with particular attention to the mixed economies in the Arctic communities. Albeit this emphasis is again appropriate and necessary, we lack good understanding of other sectors, such as cultural, knowledge and service economies. These sectors, however, play an important role in sustainable development as stand-alone entities and as parts of the mixed economy.

- **Lack of integrated understanding of socio-ecological transformation in the Arctic, of how the Arctic is transforming and the consequences thereof in and beyond the Arctic.** Despite some efforts to provide inter-regional, integrated analysis of socio-ecological transformations in the Arctic (for example, through circumpolar assessment, such as AACA, AHDR, etc.), there is still limited integration and synthesis of case studies and findings from disciplinary research.

8.3 Priorities: Agenda 2025

Selecting priorities for future research

In our selection of research priorities for the next decade we closely followed the ICARP III charge and avoided replicating science plans published in 2005 or creating extensive lists of possible research questions and approaches. We also do not aim to produce a prescriptive list of research themes. Rather, we focused on elucidating a few actionable research priorities. These priorities were selected based on several criteria: 1) the ability to address identifiable knowledge gaps; 2) the ability to build on progress over the past decade; 3) the ability to be actionable on a decadal basis; 4) the relevance to society in the sense of having practical applicability.

Agenda 2025: Research priorities

- Integrated analysis of existing and new data in longitudinal and comparative syntheses;
- New methodologies able to assess process and outcomes of sustainable development in an interdisciplinary manner (social, natural and physical sciences and humanities) and through knowledge co-production;

- Design of sustainability indicators and monitoring systems;
- Study of governance, justice, equity, legitimacy, power and agency;
- Examination of connectivities (scales, regions, time);
- Re-conceptualization of relationships between sustainability research and sustainable development practice ('trade-offs') by re-emphasizing interdependence rather than competition among the 'pillars of sustainability' within dynamic social–ecological systems.

In our opinion, the overarching research priority should be an integrated analysis of existing and new data as a part of longitudinal (back and forward) and comparative syntheses. This requires going beyond and making further use of existing case study research (both of which will constitute a move beyond case studies). We believe that the next most important methodological developments will be able to interactively assess the process and outcomes of sustainable development in an inter- and transdisciplinary manner, i.e. through involving both social and natural sciences and societal stake- and rights-holders. We should continue designing sustainability indicators that will allow us to better link social and ecological processes, developments and systems, but should also move from indicators to long-term monitoring, including the longitudinal aspect. We urge an emphasis on equity, legitimacy, power, and gender issues, and a channeling of research efforts to improve understanding of how current and changing power relations (equity, agency, gender) affect sustainability outcomes and processes. Another useful but currently understudied general approach is the study of connectivities (at different scales and how they interact in lieu of economic, cultural, ecological, atmospheric, climatic etc. processes), how they can be measured and what this means for sustainable development and sustainability in the Arctic and beyond. We suggest that we need to prioritize further inquiry into the Arctic urban sustainability, social, ecological and economic processes affecting urban sustainable development, urban-rural dynamics and interface, and regionalization of sustainability. Progress in sustainable development is not thinkable without closely considering the role of governance, e.g. role of governance in the dynamics of social–ecological systems.

Finally, in the next decade we need to carefully reconsider and possibly re-conceptualize relationships between sustainability science, which examines social-ecological systems and complexity dynamics, while sustainable development has been grounded in making progress across three traditional pillars of sustainable development (economic, environmental and social). The idea of "tradeoffs" represents one possible re-framing concept, but more work is needed to bring these approaches together.

Actionable research directions

Based on priorities outlined above and considering the recent trends in the Arctic sustainability research, as well as needs expressed by communities and other Arctic stakeholders, we also developed a list of actionable research items for the next 10 years. These items could be used by various research agencies and organizations to streamline their scholarly activities. We believe that attention to and investment in these research items will be most efficient and effective in respect to both moving the sustainability science forward and serving needs of diverse Arctic constituencies. The proposed items consist of actionable research directions, which are general suggestions on how to frame sustainability research in the Arctic, and actionable research themes, which are specific themes of research best designed to address current scholarly and practical tasks. Actionable research directions include:

- The consideration of sustainability and sustainable development in a more reflective manner. This implies considering historicity, pluralism and ethics in the use and manipulation of the concepts of 'sustainability' and 'sustainable development' in research, policy and practice. Stylized usage of the concept without critical re-evaluation may lead to further erosion of its theoretical rigor and practical benefit. A 'cookie-cutter" approach to sustainable development is not adequate, as it often precludes the engagement of different "voices" (and knowledge systems) in the process of sustainable development and may serve as an inadvertent source of marginalization and (re)colonization.
- An examination of power relationships vis-à-vis sustainable development and more attention to equity as a pillar of sustainable development. These directions will produce a much more nuanced and comprehensive understanding of sustainable development and process and outcome.
- Making greater efforts regarding the co-production of knowledge as a leading methodological framework of sustainability research. This includes fundamental, methodological and applied study of knowledge systems, methods of performing co-production and its applications to address specific sustainability challenges.
- The engagement of multi-scalar research that links scales and explores spatiotemporal dimensions of sustainability and sustainable development.

Proposed key research themes

Within these four major directions, we can list a number of priority research themes, which respond to key gaps in knowledge while providing valuable and most urgently needed contribution to theory and practice. Although the list is not exhaustive, they may serve as guidelines for prioritizing research activities and setting funding targets. These themes include:

- Synthesis of knowledge about sustainability and sustainable development in the Arctic;
- Development of integrated sustainability indicators;
- Examination of sustainable development as process: examine success stories and failures;
- Longitudinal analysis (both back and forward) of sustainable development;
- Examination of the linkages between climate change and sustainable development;
- Differentiation between development and sustainable development (for whom, what, how and why);
- Examination of the role of institutions and governance in sustainable development;
- Examination of cumulative effects of environmental and development in the context of sustainability;
- Urban areas and relationships between rural and urban;
- Role of resources as factors and instruments of sustainable development;
- Gender and youth, generational scale of sustainability;
- Building knowledge bridges with humanities;
- Communication with various stakeholders on issues of sustainability and sustainable development in the Arctic.

Note

1 Palsson, G., B. Szerszynski, S. Sörlin, J. Markes, B. Avril, C. Crumley, H. Hackmann, P. Holm, J. Ingram, A. Kirman, M. Pardo Buendía and R. Weehuizen (2013) Reconceptualizing the 'Anthropos' in the Anthropocene: integrating the social sciences and humanities in global environmental change research, *Environmental Science and Policy* 28: 3–13.

References

Aaen, S. B. (2012). *Demokratisk Legitimitet i Høringsprocesser i Forbindelse Med Storskala-Projekter i Grønland* Grønland: Grønlands Arbejdsgiverforening.

Abel, N., Cumming, D.H.M. and J.M. Anderies. (2006). Collapse and Reorganization in social–ecological systems: Questions, some ideas and policy implications. *Ecology and Society*. 11(1):17. [online] URL: http://www.ecologyandsociety.org/vol11/iss1/art17/

Abele, F. (2009). The state and the Northern social economy: Research prospects. *Northern Review*, 30, 37–56.

Abele, F. and Rodon, T. (2007). Inuit diplomacy in the global era: The strengths of multilateral internationalism. *Canadian Foreign Policy Journal*, 13(3), 45–63.

ACIA. (2004). *Impacts of a Warming Arctic: Arctic Climate Impact Assessment.* Cambridge: Cambridge University Press.

ACIA. (2005). *Arctic Climate Impact Assessment.* Cambridge: Cambridge University Press.

Adams, M., Carpenter, J., Housty, J., Neasloss, D., Paquet, P., Walkus, J. and Darimont, C. (2014). Toward increased engagement between academic and indigenous community partners in ecological research. *Ecology and Society*, 19(3).

Adger, W. N. (2000). Social and ecological resilience: are they related? *Progress in Human Geography*, 24(3), 347–364.

Adger, W. N. (2006). Vulnerability. *Global Environmental Change*, 16(3), 268–281.

AFN (Assembly of First Nations). (2009). *Ethics in First Nations Research.* http://www.afn.ca/uploads/files/rp-research_ethics_final.pdf

Agranat. G.A. (1992). *Opportunities and Realities of Developing the North.* VINTI: Moscow. [in Russian].

Airoldi, A. (2008). *The European Union and the Arctic: Policies and Actions.* Copenhagen: Nordic Council of Ministers.

Allen, J., Hopper, K., Wexler, L., Kral, M., Rasmus, S. and Nystad, K. (2014). Mapping resilience pathways of indigenous youth in five circumpolar communities. *Transcultural Psychiatry*, 51(5), 601–631.

AMAP. (1997). *Arctic Pollution Issues: A State of the Arctic Environment Report.* Oslo: Arctic Monitoring and Assessment Programme.

AMAP. (1998). *AMAP Assessment Report: Arctic Pollution Issues*. Oslo: Arctic Monitoring and Assessment Programme.

AMAP. (2002). *Arctic Pollution*. Arctic Monitoring and Assessment Programme, Oslo.

AMAP (2009). *Human Health in the Arctic*. Oslo: Arctic Monitoring and Assessment Programme.

Andersen, S. and Midttun, A. (1985). Conflict and Local Mobilization: The Alta Hydropower Project 1. *Acta Sociologica*, 28(4), 317–35. doi:10.1177/000169 938502800402.

Andrew, R. (2014). *Socio-Economic Drivers of Change in the Arctic*. Oslo: Arctic Monitoring and Assessment Programme.

Archer, C. and Scrivener, D. (2000). International co-operation in the Arctic environment, in Mark Nuttall and Terry V. Callaghan, *The Arctic: Environment, People, Policy*. Amsterdam: Overseas Publishers Association. pp.601–619.

ArcRisk. (2013). *Arctic Health Risks: Impacts on Health in the Arctic and Europe Owing to Climate-Induced Changes in Contaminant Cycling*. www.arcrisk.eu/

Arctic Council. (1996). Declaration on the Establishment of the Arctic Council. 19 September 1996. Ottawa, Canada. www.arctic-council.org.

Arctic Council. (1998). Arctic Council Terms of Reference for a Sustainable Development Program. www.arctic-council.org.

Arctic Council. (2013). *Arctic Resilience Interim Report*. Stockholm: Stockholm Environment Institute and Stockholm Resilience Centre.

Arctic Council. (2015). *The Economics of Ecosystems and Biodiversity (TEEB) for the Arctic: A Scoping Study*. Stockholm: Stockholm Environment Institute and Stockholm Resilience Centre.

Arctic Council (2016). *Arctic Resilience Report*. Stockholm: Stockholm Environment Institute and Stockholm Resilience Centre.

Arctic Governance Project. (2010). *Arctic Governance in an Era of Transformative Change: Critical Questions, Governance Principles, Ways Forward*. Report of the Arctic Governance Project. Arctic Governance Project. arcticgovernance.org.

Armitage, D., Marschke, M. and Plummer, R. (2008). Adaptive co-management and the paradox of learning. *Global Environmental Change*, 18(1), 86–98.

Armitage, D. R., Plummer, R., Berkes, F., Arthur, R. I., Charles, A. T., Davidson-Hunt, I. J. ... and McConney, P. (2009). Adaptive co-management for social–ecological complexity. *Frontiers in Ecology and the Environment*, 7(2), 95–102.

Armitage, D., Berkes, F., Dale, A., Kocho-Schellenberg, E., and Patton, E. (2011). Co-management and the co-production of knowledge: Learning to adapt in Canada's Arctic. *Global Environmental Change*, 21(3), 995–1004.

Armitage, Derek, Béné, Chris, Charles, Anthony, Johnson, Derek and Allison, Edward (2012). The interplay of well-being and resilience in applying a social–ecological perspective. *Ecology and Society*, 17(4), 15.

Armstrong, T. (1978). *Circumpolar North: Political and Economic Geography of the Arctic and Sub-Arctic*. York, UK: Methuen.

Åtland, K. (2008). Mikhail Gorbachev, the Murmansk initiative, and the desecuritization of interstate relations in the Arctic. *Cooperation and Conflict*, 43(3), 289–311.

Aull, G. H., Gabbard, P. and Timmons, J. F. (1950). United Nations Scientific Conference on the Conservation and Utilization of Resources. *Journal of Farm Economics*, 32, 95–111.

Aven, T. and Renn, O. (2012). On the risk management and risk governance of petroleum operations in the Barents Sea area. *Risk Analysis*, 32(9), 1561–1575.

Axelson, P. and Sköld, P. eds. (2011). *Indigenous Peoples and Demography: The Complex Relation between Identity and Statistics*. New York: Berghahn.

Axworthy, T. S., Koivurova, T. and Hossain, K. eds. (2012). *The Arctic Council: Its Place in the Future of Arctic Governance*. Toronto: Munk-Gordon Arctic Security Program.

Aylward, M. L. (2009). Journey to Inuuqatigiit: Curriculum development for Nunavut education. *Diaspora, Indigenous, and Minority Education*, 3(3), 137–158.

Aylward, M. L. (2010). The role of Inuit languages in Nunavut schooling: Nunavut teachers talk about bilingual education. *Canadian Journal of Education*, 33(2), 295–328.

Baev, P. (2007). *Russia's Race for the Arctic and the New Geopolitics in the North*. Jamestown Occasional Papers. Washington DC: The Jamestown Foundation. www.jamestown.org/uploads/media/Jamestown-BaevRussiaArctic_01.pdf.

Baggio, J. A., BurnSilver, S. B., Arenas, A., Magdanz, J. S., Kofinas, G. P. and De Domenico, M. (2016). Multiplex social ecological network analysis reveals how social changes affect community robustness more than resource depletion. *Proceedings of the National Academy of Sciences*, 113(48), 13708–13713.

Bals, M., Turi, A. L., Skre, I. and Kvernmo, S. (2011). The relationship between internalizing and externalizing symptoms and cultural resilience factors in Indigenous Sami youth from Arctic Norway. *International Journal of Circumpolar Health*, 70(1), 37–45.

Bankes, N. (2011). The Protection of the rights of indigenous peoples to territory through the property rights provisions of international regional human rights instruments. *The Yearbook of Polar Law Online*, 3(1), 57–112.

Bankes N. and Koivurova T. (2013). *The Proposed Nordic Saami Convention: National and International Dimensions of Indigenous Property Rights*. Oxford: Hart Publishing.

Barnhardt, R. (2005). Indigenous knowledge systems and Alaska native ways of knowing. *Anthropology & Education Quarterly*, 36(1), 8–23.

Barnhardt, R. (2010). *Creating a Place for Indigenous Knowledge in Education. Place-Based Education in the Global Age; Local Diversity*. New York: Routledge. p. 113.

Barry, T, L. Grenoble, and F. Fridriksson. (2013). Linguistic diversity. In CAFF. eds. *Arctic Biodiversity Assessment. Status and Trends in Arctic Biodiversity*. Akureyri, Iceland: CAFF. pp. 653–663

Berger, P. and Epp, J. R. (2006). Practices against culture that "work" in Nunavut schools: Problematizing two common practices. *McGill Journal of Education*, 41(1), 9.

Berger, T. R. (1977). *Northern Frontier, Northern Homeland. The Report of the Mackenzie Valley Pipeline Inquiry*. Ottawa: Supply and Services Canada.

Berger, T. R. (1985). *Village Journey: The Report of the Alaska Native Review Commission*. New York: Hill and Wang.

Berkes, F. (2009). Evolution of co-management: role of knowledge generation, bridging organizations and social learning. *Journal of Environmental Management*, 90(5), 1692–1702.

Berkes, F. (2012). *Sacred Ecology*. New York: Routledge.

Berkes, F. and C. Folke. eds. (1998). *Linking Social and Ecological Systems: Management Practices and Social Mechanisms for Building Resilience*. Cambridge: Cambridge University Press.

Berkes, F., Colding, J. and Folke, C. (2000). Rediscovery of traditional ecological knowledge as adaptive management. *Ecological Applications,* 10, 1251–1262.

Berkes, F., Coldring, J. and Folke, C. eds. (2003). *Navigating Social–Ecological Systems: Building Resilience for Complexity*. Cambridge: Cambridge University Press.

Berkman, Paul Arthur and Young, Oran, R. (2009). Governance and environmental change in the Arctic Ocean. *Science*, 324, 339–40.

Berkman, P. A. and Vylegzhanin, A. N. eds. (2013). *Environmental Security in the Arctic Ocean*. Netherlands: Springer.

Berman, M., Kofinas, G. and BurnSilver, S. (2017). Measuring community adaptive and transformative capacity in the Arctic context. In Gail Fondahl and Gary N. Wilson. eds. *Northern Sustainabilities: Understanding and Addressing Change in the Circumpolar World.,* pp. 59–75.

Berman, M., Nicolson, C., Kofinas, G., Tetlichi, J. and Martin, S. (2004). Adaptation and sustainability in a small arctic community: Results of an agent-based simulation model. *Arctic*, 57(4), 401–414.

Bjerregaard, P. and Mulvad, G. (2012). The best of two worlds: how the Greenland Board of Nutrition has handled conflicting evidence about diet and health. *International Journal of Circumpolar Health*, 71. 18588.

Blaikie, P., Cannon, T., Davis, I. and Wisner, B. (1994). *At Risk: Natural Hazards, People's Vulnerability, and Disasters*. London: Routledge.

Bliss, L. C., Heal, O. W. and Moore, J. J. (1981). *Tundra Ecosystems: A Comparative Analysis*. Cambridge: Cambridge University Press.

Bone, R. M. (2009). *The Canadian North: Issues and Challenges*. Don Mills ON: Oxford University Press.

Bossel, H. (1999). *Indicators for Sustainable Development: Theory, Method, Applications*. Winnipeg: International Institute for Sustainable Development. pp.1–124.

Brattland, C. (2014). Mapping rights in coastal Sami seascapes. *Arctic Review*, 1(1), 28–53.

Bravo, M. T. and Sörlin, S. (2002). *Narrating the Arctic. A Cultural History of Nordic Scientific Practices*. Canton, MA: Science History Publications

Brawn, J., Miller, P. C. and Tieszen, L. L. (1980). *An Arctic Ecosystem: The Coastal Tundra at Barrow, Alaska*. Stroudsburg PA: Dowden, Hutchinson and Ross.

Breum, M. (2011). *Når Isen Forsvinder*. Copenhagen: Gyldendal.

Broderstad, E. G. and Eythórsson, E. (2014). Resilient communities? Collapse and recovery of a social–ecological system in Arctic Norway. *Ecology and Society*, 19(3). doi:10.5751/ES-06533-190301.

Bronen, R. (2011). Climate-induced community relocations: creating an adaptive governance framework based in human rights doctrine. *NYU Review of Law & Social Change*, 35, 357.

Bronen, R. and Chapin III, F. S. (2013). Adaptive governance and institutional strategies for climate-induced community relocations in Alaska. *Proceedings of the National Academy of Sciences*, 110(23), 9320–9325.

Brundtland, G. (1987). *Our Common Future*. Oxford: Oxford University Press.

Brustad, M., Hansen, K. L., Broderstad, A. R., Hansen, S. and Melhus, M. (2014). A population-based study on health and living conditions in areas with mixed Sami and Norwegian settlements: The SAMINOR 2 questionnaire study. *International Journal of Circumpolar Health*, 73.

BurnSilver, Shauna, Boone, Randall, Kofinas, Gary and Brinkman, Todd (2017). Modeling tradeoffs in a rural Alaska economy: Hunting, working and sharing in the face of economic and ecological change. In Michelle Hegmon. ed. *The Give and Take of Sustainability: Archaeological and Anthropological Perspectives on Tradeoffs*. New York: Cambridge University Press.

Burton, I., Kates, R. W. and White, G. F. (1978). *The Environment as Hazard*. New York: Oxford University Press.

Burton, I., Huq, S., Lim, B., Pilifosova, O. and Schipper, E. L. (2002). From impacts assessment to adaptation priorities: The shaping of adaptation policy. *Climate Policy*, 2(2–3), 145–159.

Byers, M. (2009). *Who Owns the Arctic? Understanding Sovereignty Disputes in the North*. Vancouver BC: Douglas & Mcintyre.

Byrnea, J., Hughesa, K., Wilson, R. B. and Kurdgelashvili, L. (2007). American policy conflict in the greenhouse: Divergent trends in federal, regional, state, and local green energy and climate change policy. *Energy Policy*, 35(9), 4555–4573.

CAFF (Conservation of Arctic Flora and Fauna). (2010). *Indices and Indicators*. https://www.caff.is/indices-and-indicators

Cameron, E., Mearns, R. and McGrath, J. T. (2015). Translating climate change: Adaptation, resilience, and climate politics in Nunavut, Canada. *Annals of the Association of American Geographers*, 105(2).

Caradonna, J. L. (2014). *Sustainability: A History*. Oxford: Oxford University Press.

Carothers, C., Lew, D. K. and Sepez, J. (2010). Fishing rights and small communities: Alaska halibut IFQ transfer patterns. *Ocean and Coastal Management*, 53(9), 518–523.

Carpenter, B. (2009). Warm is the new cold: Global warming, oil, UNCLOS article 76, and how an Arctic treaty might stop a new Cold War. *Environmental Law*, 39(1).

Carpenter, S. and Cottingham, K. (1997). Resilience and restoration of lakes. *Conservation Ecology*, 1(1).

Carson, D. B. ed. (2011). *Demography at the Edge: Remote Human Populations in Developed Nations*. Farnham, UK: Ashgate.

Carson, R. (1962). *Silent Spring*. Boston MA: Houghton Mifflin.

Cavalieri, S., McGlynn, E., Stoessel, S., Stuke, F., Bruckner, M., Polzin, C. ... Nilsson, A. E. (2010). *EU Arctic Footprint and Policy Assessment*. Final Report.

Berlin: Ecologic Institute. http://arctic-footprint.eu/sites/default/files/AFPA_Final_Report.pdf

Chan, H.M., Fediuk, K., Hamilton, S., Rostas, L., Caughey, A., Kuhnlein, H., Egeland, G. and Loring, E., (2006). Food security in Nunavut, Canada: Barriers and recommendations. *International Journal of Circumpolar Health*, 65(5), 416–431.

Chance, N. and Andreeva, E. (1995). Sustainability, equity, and natural resource development in Northwest Siberia and Arctic Alaska. *Human Ecology*, 23(2), 217–240.

Chapin III, F. S., Trainor, S. F., Huntington, O., Lovecraft, A. L., Zavaleta, E., Natcher, D. C. ... and Fresco, N. (2008). Increasing wildfire in Alaska's boreal forest: pathways to potential solutions of a wicked problem. *BioScience*, 58(6), 531–540.

Chapin III, F. S., Kofinas, G. P. and Folke, C. eds. (2009). *Principles of Ecosystem Stewardship: Resilience-Based Natural Resource Management in a Changing World*. Netherlands: Springer.

Chapin III, F.S., Trainor, S.F., Cochran, P., Huntington, H., Markon, C.J., McCammon, M., McGuire, A.D. and Serreze, M. (2014). Alaska. In J.M. Melillo, T.C. Richmond, and G.W. Yohe. eds. *Climate Change Impacts in the United States: The Third National Climate Assessment*. pp. 514–536.

Chapin, F. S., Lovecraft, A. L., Zavaleta, E. S., Nelson, J., Robards, M. D., Kofinas, G. P. and Naylor, R. L. (2006). Policy strategies to address sustainability of Alaskan boreal forests in response to a directionally changing climate. *Proceedings of the National Academy of Sciences*, 103(45), 16637–16643.

Chapin, F. S., Jefferies, R. L., Reynolds, J. F., Shaver, G. R., Svoboda, J. and Chu, E. W. eds. (2012). *Arctic Ecosystems in a Changing Climate: An Ecophysiological Perspective*. San Diego CA: Academic Press.

Chapin, F. Stuart, Sommerkorn, M., Robards, M.D, and Hillmer-Pegram, K. (2015). Ecosystem stewardship: A resilience framework for Arctic conservation. *Global Environmental Change*, 34 (September), 207–17. doi:10.1016/j.gloenvcha.2015.07.003.

Chaturvedi, S. (2012). Geopolitical transformations: "Rising" Asia and the future of the Arctic Council. In T. S. Axworthy, T. Koivurova and W. Hasanat eds. *The Arctic Council: Its Place in the Future of Arctic Governance*. Toronto: Munk-Gordon Arctic Security Program, pp. 226–60.

Chatwood, S. and Bjerregaard, P. (2012). Global health – A circumpolar perspective. *American Journal of Public Health,* 102, 1246–1249.

Chen, G. (2012). China's emerging Arctic strategy. *The Polar Journal*, 2(2), 358–371.

Christensen, M. (2013). Arctic climate change and the media: The news story that was. In M. Christensen, A. E. Nilsson, and N. Wormbs. eds. *Media and the Politics of Arctic Climate Change: When the Ice Breaks*, New York: Palgrave Macmillan, pp. 26–51.

Christensen, M., Nilsson, A. E. and Wormbs, N. eds. (2013). *Media and the Politics of Arctic Climate Change: When the Ice Breaks*. New York: Palgrave Macmillan.

Christie, P. and Sommerkorn, M. (2012). *RACER: Rapid Assessment of Circum-Arctic Ecosystem Resilience*. 2nd edn. Ottowa ON: WWG Global Arctic Programme.

Christoffersen, L. E. (1997). IUCN: A Bridge-Builder for Nature Conservation. *Green Globe Yearbook of International Co-operation on Environment and Development 1997* Oslo: Fridtjof Nansen Institute. pp. 59–69.

Clark, D. and Workman, L. (2013). Transformations in subsistence systems in the southwest Yukon Territory, Canada. In The Artic Council. eds. *Arctic Resilience Interim Report 2013*. Stockholm: Stockholm Environment Institute and Stockholm Resilience Centre. pp. 105–8.

Clark, W. C. (2007). Sustainability science: A room of its own. *Proceedings of the National Academy of Sciences*, 104(6), 1737.

Clark, W.C. and Dickson, N. M. (2003). Sustainability science: The emerging research program. *Proceedings of the National Academy of Sciences*, 100(14), 8059–8061.

Coates, K. and Poelzer, G. (2010). *On the Front Lines of Canada's Northern Strategy*. Saskatoon SK: Federation of Canadian Municipalities/Fédération canadienne des municipalités.

Cochran, P., Huntington, O.H., Pungowiyi, C., Tom, S., Chapin, F.S., Huntington, H.P. ... Trainor, S.F. (2013). Indigenous frameworks for observing and responding to climate change in Alaska. *Climatic Change*, 120(3), 557–567.

Collignon, B. (2006). *Knowing Places. The Inuit, Landscapes and the Environment*. Edmonton AB: CCI Press. pp. 204.

Conwentz, H. (1914). *Über den Schutz der Natur Spitzbergens. Denkschrift überreicht der Spitzbergenkonferenz in Kristiania*. Berlin: Borntraeger.

Cornell, S., Forbes, B. C., McLennan, D., Molau, U., Nuttall, M., Overduin, P. and Wassmann, P. (2013). Thresholds in the Arctic. In The Artic Council. eds. *Arctic Resilience Interim Report 2013*. Stockholm: Stockholm Environment Institute and Stockholm Resilience Centre. pp. 37–69.

Cote, M. and Nightingale, A. J. (2012). Resilience thinking meets social theory: situating social change in socio-ecological systems (SES) research. *Progress in Human Geography*, 36(4), 475–489.

Coulthard, S. (2012). Can we be both resilient and well, and what choices do people have? Incorporating agency into the resilience debate from a fisheries perspective. *Ecology and Society*, 17(1), 4.

Crate, S. (2006). Investigating local definitions of sustainability in the Arctic: Insights from post-Soviet Sakha villages. *Arctic*, 59(3), 294–310.

Crate, S. A. (2008). "Eating Hay": The ecology, economy and culture of Viliui Sakha smallholders of Northeastern Siberia. *Human Ecology*, 36(2), 161–174.

Cruikshank, J. (2005). *Do Glaciers Listen? Local Knowledge, Colonial Encounters and Social Imagination*. Vancouver BC: UBC Press.

Cumming, G.S., Barnes, G., Perz, S., Schmink, M., Sieving, K.E., Southworth, J. ... Van Holt, T. (2005). An exploratory framework for the empirical measurement of resilience. *Ecosystems*, 8(8), 975–987.

Cunsolo Willox, A., Harper, S. L., Ford, J. D., Landman, K., Houle, K. and Edge, V. L. (2012). From this place and of this place: Climate change, sense of place, and health in Nunatsiavut, Canada. *Social Science & Medicine*, 75(3), 538–547.

Cunsolo Willox, A., Harper, S. L. and Edge, V. L. (2013a). Storytelling in a digital age: digital storytelling as an emerging narrative method for preserving and promoting indigenous oral wisdom. *Qualitative Research*, 13(2), 127–147.

Cunsolo Willox, A., Harper, S.L., Ford, J.D., Edge, V.L., Landman, K., Houle, K., Blake, S. and Wolfrey, C. (2013b). Climate change and mental health: An exploratory case study from Rigolet, Nunatsiavut, Canada. *Climatic Change*, 121(2), 255–270.

Dahl, J., Jessen, A., MacDonald J., Minde, H., Novikova, N., Nuttall, M., Petersen, S., Poppel, B., Sambo, D., Olsen C. (2005). *ICARP II Science Plan 2: Indigenous Peoples and Change in the Arctic: Adaptation, Adjustment and Empowerment*. Copenhagen, Denmark, 10–12 November 2005 (http://www.aosb.org/icarp_ii/index.html).

Dahl, J., Fondahl, G., Petrov, A. and Fjellheim, R. S. (2010). Fate control. In J. N. Larsen, P. Schweitzer and G. Fondahl.. (2010). *Arctic Social Indicators II: A Follow Up to the Arctic Human Development Report*. Copenhagen. Nordic Council of Ministers. pp. 129–146.

Daley, P. and James, B. (2004). *Cultural Politics and Mass Media: Alaska Native Voices*. Urbana IL: University of Illinois.

Dana, L. P., Etemad, H. and Wright, R. W. (2008). Toward a paradigm of symbiotic entrepreneurship. *International Journal of Entrepreneurship and Small Business*, 5(2), 109–126.

Danilov-Danilyan, V.I. (2003). Sustainable Development (Theoretical and Methodological Analysis) Ekonomika i Matematicjeskie Metody. [in Russian].

Dannevig, H., Rauken, T. and Hovelsrud, G. (2012). Implementing adaptation to climate change at the local level. *Local Environment*, 17(6–7), 597–611.

Declaration on the Protection of the Arctic Environment. (1991). *Arctic Environmental Protection Strategy*. http://www.arctic-council.org/index.php/en/document-archive/category/4-founding-documents.

Desbiens, C. (2010). Step lightly, then move forward: Exploring feminist directions for Northern research. *The Canadian Geographer/Le Géographe canadien*, 54(4), 410–416.

Dittmer, J., Moisio, S., Ingram, A. and Dodds, K. (2011). Have you heard the one about the disappearing ice? Recasting Arctic geopolitics. *Political Geography*, 30(4), 202–214.

Dodds, Klaus, Rhemann, Jennifer, Sellheim, Nikolas, Kankaanpää, Paula, Breum, Martin, Molenaar, Erik J. … Graczyk, Piotr. (2012). Various chapters in Thomas S. Axworthy, Timo Koivurova, and Waliul Hasanat. eds. *The Arctic Council: Its Place in the Future of Arctic Governance*. Toronto, Canada: Munk-Gordon Arctic Security Program.

Duarte, C.M. and P. Wassmann. eds. (2011). *Arctic Tipping Points*. Bilbao: Fundación BBVA.

Dudarev, A. A. (2012). Dietary exposure to persistent organic pollutants and metals among Inuit and Chukchi in Russian Arctic Chukotka. *International Journal of Circumpolar Health*, 71.18592.

Dudarev, A. A., Dorofeyev, V. M., Dushkina, E. V., Alloyarov, P. R., Chupakhin, V. S., Sladkova, Y. N. … Evengard, B. (2013). Food and water security issues in Russia III: Food-and waterborne diseases in the Russian Arctic, Siberia and the Far East, 2000–2011. *International Journal of Circumpolar Health*, 72(1), 21856.

Dudeck, S. (2013). Challenging the state educational system in Western Siberia: taiga school by the Tiuitiakha River. In E. Kasten and T. De Graaf. eds.

Sustaining Indigenous Knowledge: Learning Tools and Community Initiatives for Preserving Endangered Languages and Local Cultural Heritage. Fürstenberg, Germany: Kulturstiftung Sibirien. pp. 129–157.

Duhaime, G. and Édouard, R. (2015). Monetary Poverty in Inuit Nunangat. *Arctic*, 68(2), 223–232.

Duhaime, Gérard, Lemelin, André, Didyk, Vladimir, Goldsmith, Oliver, Winther, Gorm, Caron, Andrée, Bernard, Nick and Godmaire, Anne. (2004). *Economic Systems*. Akureyri Iceland: Stefansson Arctic Institute and the Arctic Council.

Dydik, V. V. and Ryabova, L. A. (2014). Single-industry towns of the Russian Arctic: development strategy on the case study of the cities/towns in the Murmansk Oblast. *Ekonomicheskie i Sotsialnye Peremeny*, (34), 84.

Eakin, H. and Bojorquez-Tapia, L. A. (2008). Insights into the composition of household vulnerability from multicriteria decision analysis. *Global Environmental Change*, 18(1), 112–127.

Eakin, H. and Luers, A. L. (2006). Assessing the vulnerability of social-environmental systems. *Annual Review of Environment and Resources*, 31(1), 365.

Ehrlich, P.R. (1968). *The Population Bomb.*, New York: Ballantine Books.

Eilmsteiner-Saxinger, G. (2011). "We feed the nation"": Benefits and challenges of simultaneous use of resident and long-distance commuting labour in Russia's Northern hydrocarbon Industry. *Journal of Contemporary Issues in Business and Government*, 17(1), 53.

Eira, I. M. G., Jaedicke, C., Magga, O. H., Maynard, N. G., Vikhamar-Schuler, D. and Mathiesen, S. D. (2013). Traditional Sámi snow terminology and physical snow classification – Two ways of knowing. *Cold Regions Science and Technology*, 85, 117–30. doi:10.1016/j.coldregions.2012.09.004.

Ellis, F. (2000). *Rural Livelihoods and Diversity in Developing Countries*. Oxford: Oxford University Press.

ELOKA. (2015). Exchange for Local Observations and Knowledge of the Arctic (ELOKA). 25 October 2015. http://eloka-arctic.org.

Elzinga, A. (2013). The Nordic nations in Polar science: Expeditions, international polar years and their geopolitical dimensions. In S. Sörlin. ed. *Science, Geopolitics and Culture in the Polar Region. Norden Beyond Borders*. Farnham, UK: Ashgate. pp. 357–91.

Emmerson, C. (2010). *The Future History of the Arctic*. New York: Public Affairs.

English, J. (2013). *Ice and Water: Politics Peoples and the Arctic Council*. Toronto: Penguin.

Fabinyi, M., Evans, L. and Foale, S. J. (2014). Social–ecological systems, social diversity, and power: Insights from anthropology and political ecology. *Ecology and Society*. 19(4), 28.

Feit, H. A. (1995). Hunting and the quest for power: The James Bay Cree and Whitemen in the 20th century. In R. B. Morrison and C.Wilson, eds. *Native Peoples: The Canadian Experience*. Toronto: Oxford University Press. pp. 181–223.

Ferris, E. (2013). *A Complex Constellation: Displacement, Climate Change and Arctic Peoples*. Brookings–LSE Project on Internal Displacement. http://www.brookings.edu/~/media/research/files/papers/2013/1/30-arctic-ferris/30-arctic-ferris-paper.pdf

Feshbach, M. and Friendly, A. (1992). *Ecocide in the USSR: Health and Nature Under Siege*. New York: Basic Books.

Filonchik, O.A. (2011). Филончик, О. А. (2011). Традиционные знания народов, проживающих на территории Приенисейского края, в области природополь зования. Современные исследования социальных проблем, 5(1).

Folke, C. (2006). Resilience: The emergence of a perspective for social–ecological systems analyses. *Global Environmental Change*, 16(3), 253–267.

Folke, C., Carpenter, S., Walker, B., Scheffer, M., Elmqvist, T., Gunderson, L. and Holling, C. S. (2004). Regime shifts, resilience, and biodiversity in ecosystem management. *Annual Review of Ecology, Evolution and Systematics*, 35, 557–581.

Fondahl, G. (1997). Environmental degradation and indigenous land claims in Russia's North. In: Smith, E. and J. McCarter. eds. *Contested Arctic*. Washington DC: University of Washington Press. pp. 68–87.

Fondahl, G., and Poelzer, G. (2003). Aboriginal land rights in Russia at the beginning of the twenty-first century. *Polar Record*, 39(02), 111–122.

Fondahl, G., Lazebnik, O., Poelzer, G. and Robbek, V. (2001). Native "land claims", Russian style. *The Canadian Geographer/Le Géographe Canadien*, 45(4), 545–561.

Forbes, B. C. (2013). Cultural resilience of social–ecological systems in the Nenets and Yamal-Nenets autonomous Okrugs, Russia: A focus on reindeer nomads of the tundra. *Ecology and Society*, 18(4), 36.

Forbes, B. C., Bölter, M., Müller-Wille, L., Hukkinen, J., Müller, F., Gunslay, N. and Konstantinov, Y. eds. (2006). Reindeer management in northernmost Europe: Linking practical and scientific knowledge in social–ecological systems. *Ecological Studies*, 184.

Forbes, B. C., Stammler, F., Kumpula, T., Meschtyb, N., Pajunen, A. and Kaarlejärvi, E. (2009). High resilience in the Yamal-Nenets social–ecological system, West Siberian Arctic, Russia. *Proceedings of the National Academy of Sciences*, 106(52), 22041–48. doi:10.1073/pnas.0908286106.

Ford, J. D. and Furgal, C. (2009). Foreword to the special issue: Climate change impacts, adaptation and vulnerability in the Arctic. *Polar Research*, 28(1), 1–9.

Ford, J. D. and Goldhar, C. (2012). Climate change vulnerability and adaptation in resource dependent communities: A case study from West Greenland. *Climate Research*, 54(2), 181–196.

Ford, J.D. and Pearce, T. (2010). What we know, do not know and need to know about climate change vulnerability in the western Canadian Arctic: A systematic literature review. *Environmental Research Letters*, 5(1), 014008.

Ford, J. D., Smit, B., and Wandel, J. (2006). Vulnerability to climate change in the Arctic: A case study from Arctic Bay, Canada. *Global Environmental Change*, 16(2), 145–160.

Ford, J. D., Smit, B., Wandel, J., Allurut, M., Shappa, K., Ittusarjuat, H., and Qrunnut, K. (2008). Climate change in the Arctic: Current and future vulnerability in two Inuit communities in Canada. *The Geographical Journal*, 174(1), 45–62.

Ford, J. D., Pearce, T., Duerden, F., Furgal, C., and Smit, B. (2010). Climate change policy responses for Canada's Inuit population: The importance of and opportunities for adaptation. *Global Environmental Change*, 20(1), 177–191.

Ford, J. D., Berrang-Ford, L. and Paterson, J. (2011). A systematic review of observed climate change adaptation in developed nations. *Climatic change*, 106(2), 327–336.

Ford, J. D. and Smit, B. (2004). A framework for assessing the vulnerability of communities in the Canadian Arctic to risks associated with climate change. *Arctic*, 389-400.

Gallopín, G. C. (2006). Linkages between vulnerability, resilience, and adaptive capacity. *Global Environmental Change*, 16(3), 293–303.

Gamble, D. J. (1978). The Berger Inquiry: An impact assessment process. *Science*, 199(3), 946–952.

Gearheard, S., Matumeak, W., Angutikjuaq, I., Maslanik, J., Huntington, H.P., Leavitt, J., Kagak, D.M., Tigullaraq, G. and Barry, R.G. (2006). "It's not that simple": A collaborative comparison of sea ice environments, their uses, observed changes, and adaptations in Barrow, Alaska, USA, and Clyde River, Nunavut, Canada. Ambio, 35(4), 203–211.

Gearheard, S., Pocernich, M., Stewart, R., Sanguya, J. and Huntington, H. P. (2010). Linking Inuit knowledge and meteorological station observations to understand changing wind patterns at Clyde River, Nunavut. *Climatic Change*, 100(2), 267–294.

Gearheard, S., Aporta, C., Aipellee, G. and O'Keefe, K. (2011). The Igliniit project: Inuit hunters document life on the trail to map and monitor arctic change. *The Canadian Geographer/Le Géographe canadien*, 55(1), 42–55.

Gjørv, G. H., Bazely, D., Goloviznina, M. and Tanentzap, A. (2013). *Environmental and Human Security in the Arctic*. Abingdon, UK: Routledge.

Glomsrød, S. and Aslaksen, I. (2006). The Economy of the North. Oslo: Statistics Norway.

Glomsrød, S. and Aslaksen, I. (2008). *The Economy of the North 2008*. Oslo: Statistics Norway.

Gordon, H. (2015). Building sustainable relationships in the Arctic: Indigenous communities and scientists. In G. Fondahl and G. Wilson.eds. *Northern Sustainabilities: Vulnerability, Resilience, and Prosperity in the Circumpolar World*. Switzerland: Springer.

Grant, Shelagh, D. (2010). *Polar Imperative – A History of Arctic Sovereignty in North America*. Vancouver, Toronto, Berkeley: Douglas and McIntyre.

Graybill, J. K. (2007). Continuity and change: Re-constructing geographies of the environment in late Soviet and post-soviet Russia. *Area*, 39(1), 6–19.

Graybill, J.K. (2009). Places and identities on Sakhalin Island: Situating the emerging movements for "Sustainable Sakhalin". In J. Agyeman and E. Ogneva-Himmelberger. eds. *Environmental Justice of the Former Soviet Union*. Boston MA: MIT Press.

Graybill, J.K. (2013a). Environmental politics on Russia's Pacific Edge: Reactions to energy development in the Russian sea of Okhotsk. *Pacific Geographies*, 40, 11–16.

Graybill, J.K. (2013b). Mapping an emotional topography of an ecological home-land: The case of Sakhalin Island, Russia. *Emotion, Space, and Society*, 8, 39–50.

Greer, A., Ng, V. and Fisman, D. (2008). Climate change and infectious diseases in North America: The road ahead. *Canadian Medical Association Journal*, 178(6), 715–722.

Grenier, A. and Müller, D. eds. (2011). *Polar Tourism: A Tool for Regional Development*. Montreal QC: Presses de l'Université du Québec.

Grenoble, Lenore, A. and Olsen, Carl, C. (2014). Language and well-being in the Arctic: Building indigenous language vitality and sustainability. In L. Heininen, H. Exner-Pirot and J. Pliuffe, (eds.) *Arctic Yearbook 2014*. Akureyri, Iceland: Northern Research Forum.

Griffiths, F., Huebert, R. and Lackenbauer, P. W. (2011). *Canada and the Changing Arctic: Sovereignty, Security, and Stewardship*. Waterloo ON: Wilfrid Laurier University Press.

Gunn, A., Russell, D. and Greig, L. (2014). Insights into integrating cumulative effects and collaborative co-management for migratory tundra caribou herds in the Northwest Territories, Canada. *Ecology and Society*, 19(4), 4.

Guston, D. H. (2001). Boundary organizations in environmental policy and science: An introduction. *Science, Technology, and Human Values*, 26(4), 399–408.

Haasnoot, M., Kwakkel, J. H., Walker, W. E. and ter Maat, J. (2013). Dynamic adaptive policy pathways: A method for crafting robust decisions for a deeply uncertain world. *Global Environmental Change*, 23(2), 485–498.

Hacquebord, L. (2012). The history of exploration and exploitation of the Atlantic Arctic and its geopolitical consequences. In: L. Hacquebord. (ed). *Lashipa: History of Large Scale Resource Exploitation in Polar Areas*. Gronigen: University of Gronigen, Arctic Centre. pp. 127–146.

Hall, C. and Saarinen, J. (2010). Tourism and Change in Polar Regions: Climate, Environment and Experience. Abingdon, UK: Routledge.

Hamilton, L. C. and Lammers, R. B. (2011). Linking pan-Arctic human and physical data. *Polar Geography*, 34(1–2), 107–123.

Hansen K.L. (2011) Ethnic Discrimination and Bullying in Relation to Self-Reported Physical and Mental Health in Sami Settlement Areas in Norway (PhD thesis). University of Tromso, Norway: The Saminor Study.

Hart, M. (2010). Colonization, social exclusion, and indigenous health. In L. Fernandez, S. Mackinnon and J. Silver. eds. *The Social Determinants of Health in Manitoba*. pp. 91–102.

Heinämäki, L. (2009). Protecting the rights of indigenous peoples: Promoting the sustainability of the global environment?. *International Community Law Review*, 11(1), 3–68.

Heininen, L. (2010). Globalization and security in the circumpolar North. In L. Heininen and C. Southcott. eds. *Globalization and the Circumpolar North*, Fairbanks AK: University of Alaska Press, pp. 221–64.

Heininen, L. (2011). The end of the post-Cold War in the Arctic. *Nordia Geographical Publications*, 40(4), 31–42.

Heininen, L. (2013). A new Northern security. In G. Hoogensen, D. Bazely, M. Goloviznina and A. Tanentzap. eds. *Environmental Change and Human Security in the Arctic*. Abingdon, UK: Earthscan (Routledge), pp. 7–57.

Heininen, L. and Nicol, H. N. (2007). The importance of Northern dimension foreign policies in the geopolitics of the circumpolar North. *Geopolitics*, 12(1), 133–165.

Heininen, L. and Southcott, C. eds. (2010). *Globalization and the Circumpolar North*. Fairbanks AK: University of Alaska Press.

Heininen, L., Exner-Pirot, H., and Pliuffe, J. eds. (2016). *Arctic Yearbook 2016*. Akureyri, Iceland: Northern Research Forum.

Heleniak, T. (2014). Arctic populations and migration. In J.N. Larsen, and G. Fondahl eds, *Arctic Human Development Report: Regional Processes and Global Linkages*. Copenhagen: Nordic Council of Ministers. pp. 53–104.

Henriksen, T. and Ulfstein, G. (2011). Maritime delimitation in the Arctic: The Barents Sea Treaty. *Ocean Development & International Law*, 42(1–2), 1–21.

Heymann, M., Knudsen, H., Lolck, M. L., Nielsen, H., Nielsen, K. H. and Ries, C. J. (2010). Exploring Greenland: Science and technology in Cold War settings. *Scientia Canadensis*, 33(2), 11–42.

Hitch, M. and Fidler, C. R. (2007). Impact and benefit agreements: A contentious issue for environmental and aboriginal justice. *Environments Journal*, 35(2), 45–69.

Holling, C. S. (2002). Resilience and adaptive cycles. In L. Gunderson, C. Holling. eds. *Panarchy, Understanding Transformations in Human and Natural Systems*. pp. 25–62.

Holm, P., Goodsitc, M. E., Cloetingh, S., Agnoletti, M., Moldan, B. Lang, D. J. ... Zondervan, R. (2013). Collaboration between the natural and human sciences. *Global Change Research, Environmental Science and Policy*, 28, 25–35.

Honneland, G. and Stokke, O. (2006). International Cooperation and Arctic Governance. Regime Effectiveness and Northern Region Building. Abingdon, UK: Routledge.

Hoogensen Gjørv, G., Bazely, D. R, Goloviznina, M. and Tanentzap, A. J. eds. (2014). *Environmental and Human Security in the Arctic*. Abingdon, UK: Routledge.

Hough, P. (2013). *International Politics of the Arctic: Coming in from the Cold*. Abingdon, UK: Routledge.

Hovelsrud, G. K. and Smit, B. eds. (2010). *Community Adaptation and Vulnerability in Arctic Regions*. Dordrecht: Springer.

Howitt, R. (2001). *Rethinking Resource Management: Justice, Sustainability and Indigenous Peoples*. Abingdon, UK: Routledge.

Huebert, R. N., Exner-Pirot, H., Lajeunesse, A. and Gulledge, J. (2012). Climate Change and International security: The Arctic as a Bellweather. Arlington, VA: Center for Climate and Energy Solutions.

Huet, C., Rosol, R. and Egeland, G. M. (2012). The prevalence of food insecurity is high and the diet quality poor in Inuit communities. *The Journal of Nutrition*, 142(3), 541–547.

Hughes, T.P., Baird, A.H., Bellwood, D.R., Card, M., Connolly, S.R., Folke, C. ... Lough, J.M. (2003). Climate change, human impacts, and the resilience of coral reefs. *Science*, 301(5635), 929–933.

Huntington, H., Crate, S. Forbers, B., Herrmann, A., Hoel, A., Huskey, L. ... Krisijansson, K. (2005). *ICARP Science Plan 1: Arctic Economies and Sustainable Development*. Copenhagen, Denmark, 10–12 November 2005. http://www.aosb.org/icarp_ii/index.html.

Huskey, L. (2006). Limits to growth: Remote regions, remote institutions. *The Annals of Regional Science*, 40(1), 147–155.

Huskey, L. (2010). Globalization and the economics of the North. In L. Heininen and C. Southcott. eds. (2010). *Globalization and the Circumpolar North.* Fairbanks AK: University of Alaska Press. pp. 57–90.

Huskey, L. and Southcott, C. (2010). *Migration in the Circumpolar North: Issues and Contexts.* Edmonton AB: CCI Press.

ICC. (2009). *Circumpolar Inuit Declaration on Sovereignty in the Arctic.* Ottowa ON: Inuit Circumpolar Council Canada.

ICC Alaska. (2015). *Alaskan Inuit Food Security Conceptual Framework: How to Assess the Arctic From an Inuit Perspective: Summary Report and Recommendations Report.* Anchorage, AK: ICC Alaska.

ICC Canada. (2012). Assessing the Vitality of Arctic Indigenous Languages. Research Development Workshops. http://arcticlanguages.com/old/Arctic-Languages-Research-Development-Workshop-Report-Nov-2012.pdf

IPCC. (1997). The Regional Impacts of Climate Change. An Assessment of Vulnerability. Summary for Policymakers. *Special Report of Working Group II.* Geneva: IPCC.

IPCC. (2001). *IPCC Third Assessment Report.* Working Group II: Impacts, Adaptation and Vulnerability. Geneva: IPCC.

IPCC. (2014). Edenhofer, O., Pichs-Madruga, R., Sokona, Y., Farahani, E., Kadner, S. and Seyboth, K. *Climate Change 2014: Mitigation of Climate Change. Contribution of Working Group III to the Fifth Assessment Report of the Intergovernmental Panel on Climate Change. Transport.* Geneva: IPCC.

Isaac, T. and Knox, A. (2004). Canadian Aboriginal Law: Creating certainty in resource development. *University of New Brunswick Law Journal*, 53, 3.

I.T.K Inuit Tapiriit Kanatami (2011). *First Canadians, Canadians First: National Strategy on Inuit Education.* Ottowa ON: National Committee on Inuit Education.

ITK and NRI. (2007). *Negotiating Research Relationships with Inuit Communities: A Guide for Researchers.* Scot Nickels, Jamal Shirley, and Gita Laidler. eds. Ottawa and Iqaluit: Inuit Tapiriit Kanatami and Nunavut Research Institute.

IUCN. (1980). World Conservation Strategy. Living Resource Conservation for Sustainable Development. IUCN–UNEP–WWF.

Jakobson, L. (2010). China prepares for an ice-free Arctic. *SIPRI Insights on Peace and Security*, 2, 1–16.

Jakobson, L. and Peng, J. (2012). *China's Arctic Aspirations.* Stockholm: Stockholm International Peace Research Institute (SIPRI).

Jarashow, M., Runnels, M. B., and Svenson, T. (2006). UNCLOS and the Arctic: The path of least resistance. *Fordham International Law Journal*, 30, 1587.

Jorgensen, Joseph G. (1990).*Oil Age Eskimos.* Oakland CA: University of California Press.

Juutilainen, S. A., Miller, R., Heikkilä, L. and Rautio, A. (2014). Structural racism and Indigenous health: What Indigenous perspectives of residential school and boarding school tell us? A case study of Canada and Finland. *The International Indigenous Policy Journal*, 5(3), 3.

Kankaanpää, P. and Young, O. R. (2012). The effectiveness of the Arctic Council. *Polar Research*, 31(0). doi:10.3402/polar.v31i0.17176.

Karlsdottir, A., L. Olsen, L. Greve Harbo, L. Jungsberg, and R. O. Rasmussen. (2017). Future regional development policy for the Nordic Arctic: Foresight analysis 2013–2016. *Nordregio Report 2017*, 1. Stockholm, Sweden: Nordregio.

Kates, R. W., Clark, W. C., Corell, R., Hall, J. M., Jaeger, C. C., Lowe, I. ... and Faucheux, S. (2001). Sustainability science. *Science*, 292(5517), 641–642.

Kates, R.W., Parris, T. M. and Leiserowitz, A. (2005). What is sustainable development? *Environment*, 47(3), 8.

Keeling, A. (2010). "Born in an atomic test tube": landscapes of cyclonic development at Uranium City, Saskatchewan. *The Canadian Geographer/Le Géographe canadien*, 54(2), 228–252.

Keil, K. and Knecht, S. eds. (2017). *Governing Arctic Change: Global Perspectives*. London: Palgrave Macmillan.

Keith, R. F. and Simon, M. (1987). Sustainable development in the Northern circumpolar North. In P. Jacobs and D. A. Munro.eds. *Conservation with Equity: Strategies for Sustainable Development*. Gland, Switzerland: IUCN. pp. 209–25.

Kelly, P. M. and Adger, W. N. (2000). Theory and practice in assessing vulnerability to climate change and facilitating adaptation. *Climatic Change*, 47(4), 325–352.

Kershaw, G. G., Castleden, H and Laroque, C. P. (2014). An argument for ethical physical geography research on indigenous landscapes in Canada. *The Canadian Geographer/Le Géographe canadien*, 58(4), 393–399.

Keskitalo, C. (2004). *Negotiating the Arctic: The Construction of an International Region*. London: Routledge.

Klein, J.K. (2004). Prospects for transdisciplinarity. *Futures*, 36(4), 515–526.

Klimat och sårbarhetsutredningen. (2007). Sverige inför klimatförändringarna: Hot eller möjligheter. *SOU 2007*, 60. Stockholm: Fritzes.

Kofinas, G. P. (2005). Caribou hunters and researchers at the co-management interface: emergent dilemmas and the dynamics of legitimacy in power sharing. *Anthropologica*, 47(2), 179–196.

Kofinas, G. P. (2009). Adaptive co-management in social–ecological governance. In F.S Chapin III, C. Folke,and Gary P. Kofinas, *Principles of Ecosystem Stewardship*. New York: Springer. pp. 77–101.

Kofinas, G., Osherenko, G., Klein, D. and Forbes, B. (2000). Research planning in the face of change: The human role in reindeer/caribou systems. *Polar Research*, 19(1), 3–21.

Kofinas, G., Forbes, B., Beach, H., Berkes, F., Berman, M., Chapin, T. ... Young, O. (2005). *ICARP II Science Plan 10: A Research Plan for the Study of rapid Change, Resilience and Vulnerability in Social–ecological Systems of the Arctic*. Copenhagen, Denmark, 10–12 November 2005 (http://www.aosb.org/icarp_ii/index.html).

Kofinas, G. P., Chapin, F. S., BurnSilver, S., Schmidt, J. I., Fresco, N. L., Kielland, K. and Rupp, T. S. (2010). Resilience of Athabascan subsistence systems to interior Alaska's changing climate. This article is one of a selection of papers from dynamics of change in Alaska's Boreal forests: Resilience and vulnerability in response to climate warming. *Canadian Journal of Forest Research*, 40(7), 1347–1359.

Kofinas, G. P., Clark, D. and Hovelsrud, G. K. (2013). Adaptive and transformative capacity. *Arctic Resilience Interim Report 2013.* Stockholm: Environment Institute and Stockholm Resilience Centre, 73–93.

Kofinas, G., Chapin, T. and Lovecraft, A. eds. (2014). Pathways of resilience in a rapidly changing Alaska. *Ecology and Society,* 19.

Koivurova, T. (2008a). Alternatives for an Arctic treaty – evaluation and a new proposal. *Review of European Community and International Environmental Law,* 17(1), 14–26. doi:10.1111/j.1467-9388.2008.00580.x.

Koivurova, T. (2008b). *The Draft Nordic Saami Convention: Nations Working Together.* SSRN eLibrary. http://papers.ssrn.com/sol3/papers.cfm?abstract_id=1860313.

Koivurova, T. (2010). Limits and possibilities of the Arctic Council in a rapidly changing scene of Arctic governance. *Polar Record,* 46(02), 146–156.

Koivurova, T. (2013). Gaps in international regulatory frameworks for the Arctic Ocean. In P. A. Berkman and A. N. Vylegzhanin. eds. *Environmental Security in the Arctic Ocean,* NATO Science for Peace and Security Series C: Environmental Security. Netherlands: Springer. pp. 39–55.

Koivurova, T. and Molenaar, E. J. (2010). *International Governance and Regulation of the Marine Arctic.* World Wide Fund for Nature (WWF). http://www.wwf.se/.

Koivurova, T. and Stepien, A. (2011). How international law has influenced the national policy and law related to indigenous peoples in the Arctic. *Waikato Law Review,* 19, 123.

Koivurova, T. and VanderZwaag, D. L. (2007). The Arctic Council at 10 years: Retrospect and prospects. *UBC Law Review,* 40(1), 121–94.

Koivurova, T., Kokko, K., Duyck, S., Sellheim, N. and Stepien, A. (2011). The present and future competence of the European Union in the Arctic. *Polar Record,* 48(04), 361–371.

Konyshev, V. and Sergunin, A. (2012). The Arctic at the crossroads of geopolitical interests. *Russian Politics & Law,* 50(2), 34–54.

Kraska, J. ed. (2011). *Arctic Security in an Age of Climate Change.* Cambridge: Cambridge University Press, Cambridge.

Krupa, M. B., Chapin III, F. S. and Lovecraft, A. L. (2014). Robustness or resilience? Managing the intersection of ecology and engineering in an urban Alaskan fishery. *Ecology and Society,* 19(2), 17.

Krupnik, I. and Jolly, D. (2002). *The Earth Is Faster Now: Indigenous Observations of Arctic Environmental Change.* Fairbanks, AK: Arctic Research Consortium of the United States.

Krupnik, I., Bravo, M., Csonka, Y., Hovelsrud-Broda, G., Müller-Wille, L., Poppel, B. ... and Sörlin, S. (2005). Social sciences and humanities in the international Polar year 2007–2008: An integrating mission. *Arctic,* 58(1), 91–97.

Krupnik, I., Aporta, C., Gearheard, S., Laidler, G. J. and Holm, L. K. (2010). *SIKU: Knowing Our Ice.* New York: Springer.

Krupnik, I., Allison, I., Bell, R., Cutler, P., Hik, D. et al. (2011). *Understanding Earth's Polar Challenges: International Polar Year 2007–2008.* http://www.cabdirect.org/abstracts/20113200943.html;jsessionid=858B6B3434E6AD6255F041422FFDD0C7.

Kruse, J.A., White, R.G., Epstein, H.E., Archie, B., Berman, M., Braund, S.R., Chapin, F.S., Charlie, J., Daniel, C.J., Eamer, J. and Flanders, N. (2004). Modeling sustainability of Arctic communities: an interdisciplinary collaboration of researchers and local knowledge holders. *Ecosystems*, 7(8), 815–828.

Kruse, J., M. Lowe, S. Haley, G. Fay, L. Hamilton, and M. Berman. (2011). Arctic observing network social indicators project: overview. *Polar Geography*, 34 (1–2), 1–8.

Kryazhkov, V.A. (2010). Korennye malochislennye narody Severa v rossiyskom prave. Moscow: Norma.

Kryazhkov, V.A. (2013). Development of Russian legislation on Northern indigenous peoples. *Arctic Review on Law and Politics*, 4(2), 140–155.

Lantto, P. (2013). The birth of a national movement: The Sami and the Swedish state, 1900–1950. In K. Andersson. eds. *L'image du Sápmi II: Études comparées*, Sweden: Örebro University.

Larsen, J, and Fondahl, G. (2014). *Arctic Human Development Report: Regional Processes and Global Linkages*. Copenhagen: Nordic Council of Ministers.

Larsen, C. V. L., Pedersen, C. P., Berthelsen, S. W. and Chew, C. (2010). *Hope and Resilience: Suicide Prevention in the Arctic, November 7–8, 2009 Conference Report*. Nuuk: Greenland: Organizing Committee. https://oaarchive.arctic-council.org/handle/11374/41 on May 5, 2017.

Larsen, J. N., Schweitzer, P. and Fondahl, G. (2010). *Arctic Social Indicators II – A Follow Up to the Arctic Human Development Report*. Copenhagen: Nordic Council of Ministers.

Larsen, J. N., Schweitzer, P. and Petrov, A. eds. (2015). *Arctic Social Indicators: ASI II: Implementation.* Copenhagen: Nordic Council of Ministers.

Laruelle, M. (2013). *Russia's Arctic Strategies and the Future of the Far North*. New York: M.E. Sharpe.

Laruelle, M. ed. (2016). *New Mobilities and Social Changes in Russia's Arctic Regions*. Abingdon, UK: Routledge.

Lavrillier, A. (2013). Climate change among nomadic and settled Tungus of Siberia: continuity and changes in economic and ritual relationships with the natural environment. *Polar Record*, 49(03), 260–271.

Lazhentsev, V. N. (2005). *Sever kak ob'ekt kompleksnykh regional'nykh issledovaniy. [The North as the Object of Comprehensive Regional Studies]*. Syktyvkar, Russia: Nauchnyy sovet RAN po voprosam regional'nogo razvitiya, Komi nauchnyy tsentr UrO RAN [Scientific Council for Regional Development, Komi Scientific Center of the Ural Branch of the Russian Academy of Sciences].

Leach, M., Scoones, I. and Stirling, A., (2010). *Dynamic Sustainabilities: Technology, Environment, Social Justice*. London: Earthscan,

Levin, S. A. (1998). Ecosystems and the biosphere as complex adaptive systems. *Ecosystems*, 1(5), 431–436.

Lidestav, G., Poudyal, M., Holmgren, E. and Keskitalo, C. (2013). Shareholder perceptions of individual and common benefits in Swedish forest commons. *International Journal of the Commons*, 7(1).

Lindroth, Marjo and Sinevaara-Niskanen, Heidi. (2013). At the crossroads of autonomy and essentialism: Indigenous peoples in international environmental politics. *International Political Sociology*, 7(3), 275–293. DOI: 10.1111/ips.12023.

Linnér, B. O. (2003). *The Return of Malthus: Environmentalism and Post-War Population–Resource Crises*. Isle of Harris, Scotland: White Horse Press.

Liverman, D. M. (2001). Vulnerability to drought and climate change in Mexico. In Jeanne X. Kasperson and Roger E. Kasperson, Tokyo and New York: United Nations University Press. Global Environmental Risk, 158–77.

Loring, P. A. and Gerlach, S. C. (2009). Food, culture, and human health in Alaska: An integrative health approach to food security. *Environmental Science & Policy*, 12(4), 466–478.

Louafi, S. (2007). *Epistemic Community and International Governance of Biological Diversity: A Reinterpretation of the Role of IUCN*. Participation for Sustainability in Trade. Abingdon, UK: Ashgate. pp. 111–20.

Loukacheva, N. (2007). *The Arctic Promise: Legal and Political Autonomy of Greenland and Nunavut*. Ottowa ON: University of Toronto Press.

Loukacheva, N. (2009). Arctic indigenous peoples' internationalism: In search of a legal justification. *Polar Record*, 45(1), 51–58.

Loukacheva, N. ed. (2010). *Polar Law Textbook*. Copenhagen: Nordic Council of Ministers.

Lovecraft, A. L. and Eicken, H. eds. (2011). *North by 2020: Perspectives on Alaska's Changing Social-Ecological Systems*. Fairbanks, AK: University of Alaska Press.

Malthus, T.R. (1798). *An Essay on the Principle of Population, as it Affects the Future Improvement of Society with Remarks on the Speculations of Mr. Godwin, M. Condorcet, and Other Writers*. London: J. Johnson.

Manchuk, V. T. and Nadtochiy, L. A. (2010). The state and tendencies of the health formation in people born in North and Siberia. Byulleten'SO RAMN [*Bulletin of Siberian Branch of Russian Academy of Medical Sciences*], 30, 24–32.

Manicom, J. and Lackenbauer, P. W. (2013). East Asian states, the Arctic Council and international relations in the Arctic. CIGI *Policy Brief*, April 2013 (26). http://www.cigionline.org/sites/default/files/no26.pdf

Mathiesen, S. D., Alfthan, B., Corell, R. W., Eira, R. B. M., Eira, I. M. G., et al. (2013). Strategies to enhance the resilience of Sámi reindeer husbandry to rapid change in the Arctic. *Arctic Resilience Interim Report 2013*, Stockholm: Stockholm Environment Institute and Stockholm Resilience Centre, 109–11.

May, P. J., Jones, B. D., Beem, B. E., Neff-Sharum, E. A. and Poague, M. K. (2005). Policy coherence and component-driven policymaking: Arctic policy in Canada and the United States. *Policy Studies Journal*, 33(1), 37–63.

Mayne, J. B. (1947). FAO – The Task. *Review of Marketing and Agricultural Economics*, 15(12), 450.

Meek, C. L. (2011). Putting the US polar bear debate into context: The disconnect between old policy and new problems. *Marine Policy*, 35(4), 430–439.

Meltofte, H., Barry, T., Berteaux, D., Bültmann, H., Christiansen, J.S., Cook, J.A., Dahlberg, A., Daniëls, F.J., Ehrich, D., Fjeldså, J. and Friðriksson, F. (2013). *Arctic Biodiversity Assessment. Synthesis*. Akureyri, Iceland: Conservation of Arctic Flora and Fauna (CAFF).

McCannon, J. (2012). *A History of the Arctic: Nature, Exploration and Exploitation*. London: Reaktion Books.

McCarthy, J. J. (2001). *Climate Change 2001: Impacts, Adaptation, and Vulnerability.* Cambridge: Cambridge University Press.

McCarthy, J., Martello, M. L., Corell, R., Selin, N. E., Fox, S., et al. (2005). *Climate Change in the Context of Multiple Stressors and Resilience.* Arctic Climate Impact Assessment. New York: Cambridge University Press.

McCormick, J. (1991). *Reclaiming Paradise: The Global Environmental Movement.* Bloomington, IN: Indiana University Press.

McGregor, D. (2004). Coming full circle: Indigenous knowledge, environment, and our future. *The American Indian Quarterly*, 28(3), 385–410.

McGregor, D. (2012). Traditional knowledge: Considerations for protecting water in Ontario. *International Indigenous Policy Journal*, 3(3).

McGregor, D., Bayha, W. and Simmons, D. (2010). "Our responsibility to keep the land alive": Voices of Northern Indigenous researchers. *Pimatisiwin: A Journal of Aboriginal and Indigenous Community Health*, 8(1), 101–123.

McGregor, H. E. (2012). Curriculum change in Nunavut: Towards Inuit Qaujimajatuqangit. *McGill Journal of Education/Revue des sciences de l'éducation de McGill*, 47(3), 285–302.

McGuire, A. D., Chapin III, F. S., Walsh, J. E., and Wirth, C. (2006). Integrated regional changes in Arctic climate feedbacks: Implications for the global climate system. *Annual Review of Environment and Resources*, 31, 61–91.

MEA (Millennium Ecosystem Assessment). (2005). *Ecosystems and Human Well-Being: Biodiversity Synthesis.* Washington DC: World Resources Institute.

Melillo, J. M., Richmond, T. C. and Yohe, G. W. (2014). *Climate Change Impacts in the United States: The Third National Climate Assessment.* Washington DC: US Global Change Research Program, p. 841.

Metcalf, V. and Robards, M. (2008). Sustaining a Healthy Human–Walrus Relationship in a Dynamic Environment: Challenges for Co-Management. *Ecological Applications*, 18, S148-S156.

Miljøverndepartementet, (2010). Tilpassing til eit klima i endring: Samfunnet si sårbarheit og behov for tilpassing til konsekvensar av klimaendringane [Adaptation to a Changing Climate: Society's Vulnerability and Need for Adaptation to Consequences of Climate Change]. *Noregs offentlege utgreiingar*. 10.

Mitchell, D. (2001). *Take My Land, Take My Life: The Story of Congress's Historic Settlement of Alaska Native Land Claims, 1960-1971.* Fairbanks AK: University of Alaska Press.

Mitchell, R. B., Clark, W. C., Cash, D. W. and Dickson, N. M. eds. (2006). *Global Environmental Assessments: Information and Influence.* Boston MA: MIT Press.

Moiseev, N.N. (2000). Моисеев, Н.Н. (2000). Судьба цивилизации. Путь разума / Н. Н. Моисеев. — Москва: Яз. рус. культуры. — 223, [1] с. — (Язык. Семиотика. Культура.).

Molenaar, Erik J. (2009). Arctic fisheries conservation and management: Initial steps of reform of the international legal framework. In G. Alfredsson, Timo Koivurova, and D. K. Leary. eds. *Yearbook of Polar Law*, Leiden, Netherlands: Martinus Nijhoff Publishers.

Molenaar, Erik J. (2012a). Arctic fisheries and international law: Gaps and options to address them. *Carbon and Climate Law Review*, (1), 63–77.

Molenaar, Erik J. (2012b). Fisheries regulation in the maritime zones of Svalbard. *The International Journal of Marine and Coastal Law*, 27(1), 3–58.

Molenaar, Erik J. (2012c). Current and prospective roles of the Arctic Council system within the context of the law of the Sea. In T. S. Axworthy, T. Koivurova and K. Hossain. eds. *The Arctic Council: Its Place in the Future of Arctic Governance*. Toronto: Munk-Gordon Arctic Security Program.

Molenaar, Erik J. and Corell, R. (2009b). *Arctic Shipping*. Berlin: Ecologic Institute.

Molenaar, Erik J. and Corell, R. (2009a). *Arctic Fisheries*. Berlin: Ecologic Institute.

Müller, D. (2015). Issues in arctic tourism. In B. Evengård, J.N. Larsen and Ø. Paashe. eds. *The New Arctic*. Dordrecht: Springer. pp. 147–158.

Müller, D., L. Lundmark and R. Lemelin, eds. (2013). *New Issues in Polar Tourism: Communities, Environments, and Politics*. Dordrecht: Springer.

Murray, R. W. and Nuttall, A. D. eds. (2014). *International Relations and the Arctic: Understanding Policy and Governance*. Amhurst NY:Cambria Press.

Mustonen, T. and Mustonen, K. (2016). *Life in the Cyclic World: A Compendium of Traditional Knowledge from the Eurasian North*. Snowchange Cooperative. http://www.snowchange.org/pages/wp-content/uploads/2016/05/life-in-the-cyclic-world-final-250516.pdf.

Mutua, K. and Swaderner, B. B. eds. (2004). *Decolonizing Research in Cross-Cultural Contexts: Critical Personal Narratives*. Albany: SUNY Press.

Nabok I.L. (2014). *Realnost Etnosa*. (Reality of Ethnos). St. Petersburg: Herzen University Press. [in Russian].

Nadasdy, P. (1999). The politics of TEK: Power and the" integration" of knowledge. *Arctic Anthropology*, 36(1/2), 1–18.

Nadasdy, P. (2003). *Hunters and Bureaucrats - Power, Knowledge, and Aboriginal–State Relations in the Southwest Yukon*. Vancouver BC: UBC Press.

Natcher, D. C. (2013). Gender and resource co-management in Northern Canada. *Arctic*, 66(2), 218–221.

National Academies (2014). *The Arctic in the Anthropocene: Emerging Research Questions*. Washington DC: NAS.

Nelson, R. K. (1973). *Hunters of the Northern Forest: Designs for Survival Among the Alaska Kutchin*. Chicago IL: University of Chicago Press.

Nelson, R. K. (1983). Make Prayers to the Raven: A Koyukon View of the Northern Forest, Chicago IL: University of Chicago Press.

Neumann, A. (2010). *The EU – A Relevant Actor in the Field of Climate Change in Respect to the Arctic?* Berlin: German Institute for International and Security Affairs (SWP). http://www.swp-berlin.org/fileadmin/contents/products/arbeits papiere/Nma_WP_2010_03_ks.pdf

Nilsson, A. E. (2009a). A changing Arctic climate: Science and policy in the Arctic climate impact assessment. In T. Koivurova, C. Keskitalo and N. Banks. Eds. *Climate Governance in the Arctic*. Berlin: Springer, pp. 77–95.

Nilsson, A. E. (2009b). Arctic climate change: North American actors in circumpolar knowledge production and policy making. In H. Selin, and S. D. VanDeveer. eds. *Changing Climates in North American Politics: Institutions, Policy Making and Multilevel Governance*. Cambridge, MA: MIT Press (forthcoming). pp. 199–217.

Nilsson, A. E. (2012). The Arctic environment - from low to high politics. *Arctic Yearbook 2012*. Akureyri, Iceland: Northern Research Forum.

Nilsson, A. E. and Döscher, R. (2013). Signals from a noisy region. In M. Christensen, A. Nilsson, N. Wormbs. eds. *Media and the Politics of Arctic Climate Change: When the Ice Breaks*. p. 93.

Nilsson, A. E. and C. L. Meek. (2016). Learning to live with change. In Marcus Carson and Garry Peterson. eds. *Arctic Resilience Report*. Stockholm: Stockholm Environment Institute and Stockholm Resilience Centre.

Nilsson, A.E and Koivurova, T. (2016). Transformational change and regime shifts in the circumpolar Arctic. *Arctic Review on Law and Politics*, 7(2). doi:10.17585/arctic.v7.532.

Nilsson, A., Larsen, J. N. and Einarsson, N. (2004). *Arctic Human Development Report*. Akureyri, Iceland: Stefansson Arctic Institute.

Nilsson, A., Carlsen, H. and van der Watt, L.M. (2015). *Uncertain Futures: The Changing Global Context of the European Arctic*. SEI Working Paper. Stockholm: Stockholm Environment Institute. http://www.sei-international.org/publications?pid=2833.

Nilsson, L. M. and Evengård, B. (2013). *Food and Water Security Indicators in an Arctic Health Context*. A Report by the AHHEG/SDWG, and the AMAP/HHAG during the Swedish Chairmanship of the Arctic Council 2011–2013. http://umu.diva-portal.org/smash/record.jsf?searchId=1&pid=diva2:585006.

Nilsson, L.M., Destouni, G., Berner, J., Dudarev, A.A., Mulvad, G., Odland, J.O., Parkinson, A., Tikhonov, C., Rautio, A. and Evengård, B. (2013). A call for urgent monitoring of food and water security based on relevant indicators for the Arctic. *Ambio*, 42(7), 816–822.

Nord, D. C. (2016). *The Changing Arctic*. New York: Palgrave Macmillan. http://link.springer.com/10.1057/9781137501868

Novikova, N. (2008). Lyudy Severa: Prava na Resursy i ekspertiza. Moscow: Strategiya.

Novikova, N. (2010). The status of indigenous peoples of the Russian North, in the context of legal pluralism. *Indigenous Affairs*, 1–2/10.

Novikova, N. I. (2016). Who is responsible for the Russian Arctic? Co-operation between indigenous peoples and industrial companies in the context of legal pluralism. *Energy Research & Social Science*, 16, 98–110.

Novikova N.I. and Stepanov V.V. (2010). Indikatory kachestva zhizni korennykh malochislennykh narodov Severa Sakhalinskoi oblasti [Markers of the quality of life of the indigenous peoples of the North]. *Issledovaniya po Prikladnoi i Neotlozhnoi Etnologii*, 217.

NRC (National Research Council). (1999). *Our Common Journey: A Transition Toward Sustainability*. Washington DC: National Academies Press.

NRC (National Research Council). (2003). Cumulative environmental effects of oil and gas activities in Alaska's North Slope. Washington DC: National Academies Press.

Nuttall, M. (2010). *Pipeline Dreams: People, Environment, and the Arctic Energy Frontier*. Copenhagen: IWGIA.

Nuttall, M. (2012). Tipping points and the human world: living with change and thinking about the future. *Ambio*, 41(1), 96–105. doi:10.1007/s13280-011-0228-3.

Nuttall, M. (2013). Zero-tolerance, uranium and Greenland's mining future. *The Polar Journal*, 3(2), 368–383.

Nuttall, M., Berkes, F., Forbes, B., Kofinas, G., Vlassova, T. and Wenzel, G. (2005). Hunting, herding, fishing and gathering: Indigenous peoples and renewable resource use in the Arctic. In the ACIA Scientific Report, *Arctic Climate Impact Assessment*. pp. 649–690.

Nystad, K., Spein, A. R. and Ingstad, B. (2014). Community resilience factors among indigenous Sámi adolescents: A qualitative study in Northern Norway. *Transcultural Psychiatry*, 51(5), 651–72. doi: 10.1177/1363461514532511.

O'Brien, K., Eriksen, S. E., Schjolden, A. and Nygaard, L. P. (2004). *What's in a Word? Conflicting Interpretations of Vulnerability in Climate Change Research*. Oslo: CICERO.

O'Neill, D. (1994). *The Firecracker Boys*. New York: St. Martin's Griffin.

Oddsdóttir, E.E., Sigurðsson, A.M. and Svandal, S. eds. (2015). *Gender Equality in the Arctic. Current Realities, Future Challenges*. Reykjavik, Iceland: Ministry for Foreign Affairs, Iceland.

Olsen, C.C. (1999). Language and sustainable development. In H. Petersen and B. Poppel. eds. *Dependency, Autonomy, Sustainability in the Arctic*. Farnham, UK: Ashgate. pp. 253–258.

Olsson, P., Folke, C. and Berkes, F. (2004). Adaptive comanagement for building resilience in social–ecological systems. *Environmental Management*, 34(1), 75–90.

Oort, B. van, Bjørkan, M. and Klyuchnikova, E. M. (2015). *Future Narratives for Two Locations in the Barents Region*. CICERO Report 2015:06. Oslo: CICERO Center for International Climate and Environmental Research. http://hdl.handle.net/11250/2367371.

Oskal, A., Turi, J. M., Mathiesen, S. D. and Burgess, P. (2009). *EALÁT. Reindeer Herders Voice: Reindeer Herding, Traditional Knowledge and Adaptation to Climate Change and Loss of Grazing Lands*. Kautokeino / Guovdageadnu, Norway: International Centre for Reindeer Husbandry.

Ozkan, U. R. and Schott, S. (2013). Sustainable development and capabilities for the polar region. *Social Indicators Research*, 114(3), 1259–1283.

Paglia, E. (2016). *The Northward Course of the Anthropocene. Transformation, Temporality and Telecoupling in a Time of Environmental Crisis*. Stockholm: KTH Royal Institute of Technology. pp. 17–29

Paine, R. (1982). *Dam a River, Damn a People? Saami (Lapp) Livelihood and the Alta/Kautokeino Hydro-electric Project and the Norwegian Parliament*. Copenhagen: International Work Group for Indigenous Affairs.

Palsson, G., Szerszynski, B., Sörlin, S., Markes, J., Avril, B., Crumley, C. … Weehuizen, R. (2013). Reconceptualizing the "Anthropos" in the Anthropocene: integrating the social sciences and humanities in global environmental change research. *Environmental Science and Policy*, 28, 3–13.

Paschen, J. A. and Ison, R. (2014). Narrative research in climate change adaptation: Exploring a complementary paradigm for research and governance. *Research Policy*, 43(6), 1083–92. doi:10.1016/j.respol.2013.12.006.

Patrick, D. (2005). Language rights in Indigenous communities: The case of the Inuit of Arctic Québec. *Journal of Sociolinguistics*, 9(3), 369–389.

Pearce, T., Wright, H., Notaina, R., Kudlak, A., Smit, B., Ford, J. and Furgal, C. (2011). Transmission of environmental knowledge and land skills among Inuit men in Ulukhaktok, Northwest Territories, Canada. *Human Ecology*, 39(3), 271–288.

Pearce, T. D., Ford, J. D., Laidler, G. J., Smit, B., Duerden, F., Allarut, M. ... Goose, A. (2009). Community collaboration and climate change research in the Canadian Arctic. *Polar Research*, 28(1), 10–27.

Pelyasov, A., Galtseva, N., Batsaev, I. and Golubenko, I. (2011). Knowledge Transfer inside the Regional Economic System: The Case of Eighty Years of Economic History of the Russian North-East. Proceedings of the 51st Congress of the European Regional Science Association: "New Challenges for European Regions and Urban Areas in a Globalised World", 30 August–3 September 2011, Barcelona, Spain.

Peters, E. J. (1999). Native people and the environmental regime in the James Bay and Northern Quebec Agreement. *Arctic*, 52(4), 395–410.

Peterson, G. and Carlos Rocha, J. C. (2016). Arctic Regime Shifts and Resilience. In Marcus Carson and Garry Peterson. eds. *Arctic Resilience Report*. Stockholm: Stockholm Environment Institute and Stockholm Resilience Centre. https://www.sei-international.org/mediamanager/documents/Publications/ArcticResilienceReport-2016.pdf.

Petrov, A. (2008). Talent in the cold? Creative capital and the economic future of the Canadian North. *Arctic*, 61(2), 162–176.

Petrov, A. (2010). Post-staple bust: modeling economic effects of mine closures and post-mine demographic shifts in an arctic economy (Yukon). *Polar Geography*, 33(1-2), 39–61.

Petrov, A. (2012). Redrawing the margin: re-examining regional multichotomies and conditions of marginality in Canada, Russia and their Northern frontiers. *Regional Studies*, 46(1), 59–81.

Petrov, A. (2016). Exploring the Arctic's "other economies": knowledge, creativity and the new frontier. *The Polar Journal*, 6(1), 51–68.

Petrov, A. N., BurnSilver, S., Chapin III, F. S., Fondahl, G., Graybill, J., Keil, K., Riedlsperger, R and Schweitzer, P. (2016). Arctic sustainability research: toward a new agenda. *Polar Geography*, 39(3), 165–178.

Pintér, L., Hardi, P., Bartelmus, P. (2005). *Sustainable Development Indicators: Proposals for a Way Forward*. New York: UN Division for Sustainable Development.

Poddubikov, V.V. (2012). Поддубиков, В. В. (2012). Коренные народы на пути устойчивого развития: традиционное природопользование и проблемы сохранения природно-культурного наследия (опыт Алтае-Саянского экорегиона).Современные исследования социальных проблем, (3).

Poeltzer, G. and Wilson G.N. (2014). Governance in the Arctic: Political systems and geopolitics. In Larsen, J. N. and Fondahö, G. eds. *Arctic Human Development Report. Regional Processes and Global Linkages*. Copenhagen: Nordic Council of Ministers.

Pohl, C., Rist, S., Zimmermann, A., Fry, P., Gurung, G. S., Schneider, F. and Hadorn, G. H. (2010). Researchers' roles in knowledge co-production: experience from sustainability research in Kenya, Switzerland, Bolivia and Nepal. *Science and Public Policy*, 37(4), 267.

Poppel, B. and Kruse, J. (2009). The importance of a mixed cash-and harvest herding based economy to living in the Arctic: An analysis on the survey of living conditions in the Arctic (SLiCA). In Valerie Møller and Denis Huschka. eds. *Quality of Life and the Millennium Challenge*. Netherlands: Springer. pp. 27–42.

Poppel, B., Kruse, J., Duhaime, G., Abryutina, L. and by Marg, D. C. (2007). Survey of living conditions in the Arctic: Results. Anchorage, AK: Institute of Social and Economic Research, University of Alaska. http://www. arcticliving conditions.org.

Pryde, P. (1991). *Environmental Management in the Soviet Union*. Cambridge: Cambridge University Press.

Pulsifer, P., Gearheard, S., Huntington, H. P., Parsons, M. A., McNeave, C. and McCann, H. S. (2012). The role of data management in engaging communities in Arctic research: overview of the Exchange for Local Observations and Knowledge of the Arctic (ELOKA). *Polar Geography*, 35(3–4), 271–290.

Rabe, B. G. (2004). North American federalism and climate change policy: American state and Canadian provincial policy development. *Widener Law Journal*, 14, 121.

Ramadier, T. (2004). Transdisciplinarity and its challenges: The case of urban studies. *Futures*, 36(4), 423–439.

Rasmus, S. M. (2008). Indigenous Emotional Economies in Alaska: Surviving Youth in the Village. PhD Dissertation. Fairbanks, AK: University of Alaska.

Rasmus, S. M. (2014). Indigenizing CBPR: Evaluation of a community-based and participatory research process implementation of the Elluam Tungiinun (towards wellness) program in Alaska. *American Journal of Community Psychology*, 54(1–2), 170–179.

Rasmussen, R. O. (2014). Gender and generation: Perspectives on ongoing social and environmental changes in the Arctic. *Signs*, 40(1).

Rautio, A., Poppel, B. and Young, K. (2014). Human health and well-being. In J. Larsen and G. Fondahl (2014). *Arctic Human Development Report: Regional Processes and Global Linkages*. Copenhagen. Nordic Council of Ministers. 299–348.

Rea, K. J. (1968). *The Political Economy of the Canadian North: An Interpretation of the Course of Development in the Northern Territories of Canada to the Early 1960s*. Toronto: University of Toronto Press.

Reed, M. S. (2008). Stakeholder participation for environmental management: A literature review. *Biological Conservation*, 141(10), 2417–2431.

Richardson, B. (1977). *Strangers Devour the Land: The Cree Hunters of the James Bay Area Versus Premier Bourassa and the James Bay Development Corporation*. Toronto: Macmillan.

Riedlsperger, R. (2014). Vulnerability to Changes in Winter Trails and Travelling: A Case Study from Nunatsiavut. Master's thesis, Memorial University of Newfoundland.

Riedlsperger, R., Goldhar, C., Sheldon, T. and Bell, T. (2017). Meaning and means of "Sustainability": An example from the Inuit settlement region of Nunatsiavut, Northern Labrador. In Gail Fondahl and Gary N Wilson. eds. *Northern Sustainabilities: Understanding and Addressing Change in the Circumpolar World*. Switzerland: Springer. pp. 317–336.

Rigney, L. I. (1999). Internationalization of an Indigenous anticolonial cultural critique of research methodologies: A guide to Indigenist research methodology and its principles. *Wicazo Sa Review*, 14(2), 109–121.

Riseth, J. Å., Tømmervik, H., Helander-Renvall, E., Labba, N., Johansson, C., Malnes, E. ... Callaghan, T. V. (2011). Sámi traditional ecological knowledge as a guide to science: Snow, ice and reindeer pasture facing climate change. *Polar Record*, 47(03), 202–217.

Rittel, H.W. and Webber, M.M. (1973). Dilemmas in a general theory of planning. *Policy Sciences*, 4, 155–169.

Rodon, T. and Lévesque, F. (2015). Understanding the social and economic impacts of mining development in Inuit communities: Experiences with past and present mines in Inuit Nunangat. *Northern Review*, (41), 13.

Røseth, T. (2014). Russia's China policy in the Arctic. *Strategic Analysis*, 38(6), 841–859.

Russell, Bruce A. (1996). The Arctic environmental protection strategy and the new Arctic Council. *Arctic Research of the United States* (10), 2–8.

Sabin, P. (1995). Voices from the hydrocarbon frontier: Canada's Mackenzie Valley pipeline inquiry (1974–1977). *Environmental History Review*, 19(1), 17–48.

Saku, J. C. (2002). Modern land claim agreements and Northern Canadian Aboriginal communities. *World Development*, 30(1), 141–151.

Sale, Richard, and Potapov, Eugene. (2010). *The Scramble for the Arctic: Ownership, Exploitation and Conflict in the Far North*. London: Frances Lincoln.

Salokangas, R. and Parlee, B. (2009). The influence of family history on learning opportunities of Inuvialuit youth. *Études/Inuit/Studies*, 33 (1–2), 191–207.

Saxinger, G. (2016). Lured by oil and gas: Labour mobility, multi-locality and negotiating normality and extreme in the Russian Far North. *The Extractive Industries and Society*, 3(1), 50–59.

Scheffer, M., Carpenter, S.R., Lenton, T.M., Bascompte, J., Brock, W., Dakos, V. ... Pascual, M. (2012). Anticipating critical transitions. *Science*, 338(6105), 344–348.

Schellnhuber, H. J. (2004). *Earth System: Analysis for Sustainability*. Cambridge MA: MIT Press.

Schweitzer, P, Sköld, P. and Ulturgasheva, O. (2014). Cultures and identities. In J.N. Larsen, G. Fondahl, H. Rasmussen. eds. *Arctic Human Development Report II: Regional Processes and Global Linkages*. Copenhagen: Nordic Council of Ministers. pp. 105–150.

Scoones, I. (1998). Sustainable rural livelihoods: A framework for analysis. *IDS Working Paper 72*. Brghton, UK: Institute of Development Studies.

Selin, H. and Linnér, Björn-Ola. (2005). The Quest for Global Sustainability: International Efforts on Linking Environment and Development. CID Graduate Student and Postdoctoral Fellow Working Paper No. 5. Cambridge, MA. Science,

Environment and Development Group, Center for International Development, Harvard University. http://www.hks.harvard.edu/var/ezp_site/storage/fckeditor/file/pdfs/centers-programs/centers/cid/publications/student-fellows/wp/005.pdf.

Sen, A. (1981). *Poverty and famines: an essay on entitlement and deprivation.* Oxford: Oxford University Press.

Shadian, J. (2010). From states to polities: Reconceptualising sovereignty through Inuit governance. *European Journal of International Relations*,16(3), 485–510.

Shadian, J. M. (2014). *The Politics of Arctic Sovereignty: Oil, Ice, and Inuit Governance.* London: Routledge.

Shadian, J.M. and Tennberg, M. eds. (2009). *Legacies and Change in Polar Sciences. Historical, Legal and Political Reflections on the International Polar Year.* Farnham: Ashgate.

Sharakhmatova, V. (2011). *Observations of Climate Change by Kamchatka Indigenous Peoples.* Report for Lach Ethnoecological Information Center. Petropavlovsk-Kamchatsky: Kamchat Press.

Sinevaara-Niskanen, Heidi. (2012). Gender, Economy and Development in the North. In Monica Tennberg. ed. *Politics of Development in the Barents Region.* Rovaniemi: Lapland University Press. pp. 362–377.

Smit, B. and Pilifosova, O. (2001). Adaptation to climate change in the context of sustainable development and equity. In J. J. McCarthy, O. F. Canziani, N. A. Leary, D. J. Dokken and K. S. White. eds. *Climate Change 2001: Impacts, Adaptation, and Vulnerability - Contribution of Working Group II to the Third Assessment Report of the Intergovernmental Panel on Climate Change.* Cambridge: Cambridge University Press.

Smit, B. and Wandel, J. (2006). Adaptation, adaptive capacity and vulnerability. *Global Environmental Change*, 16(3), 282–292.

Smit, B., Hovelsrud, G. K. and Wandel, J. eds. (2008). *Community Adaptation and Vulnerability in Arctic Regions.* Heidelberg: Springer.

Sokolova, Z.P. and Stepanov, V.V. (2007). Korennye malochislennye narody Severa. Dinamika chislennosti po dannym perepisey naseleniya. *Etnograficheskie obozrenie*, 5, 75–95.

Sommerkorn, M. and Nilsson, A. (2015). Governance of Arctic ecosystem services. In CAFF, *The Economics of Biodiversity and Ecosystem Services TEEB Scoping Study for the Arctic.* Akureyri, Iceland: Conservation of Arctic Flora and Fauna. pp. 51–76.

Sörlin, S. (1988). *Framtidslandet: debatten om Norrland och naturresurserna under det industriella genombrottet.* Stockholm: Carlssons.

Sörlin, S. ed. (2013). *Science, Geopolitics and Culture in the Polar Region: Norden Beyond Borders.* Burlington VT, Farnham: Ashgate.

Sosa, I. and Keenan, K. (2001). *Impact Benefit Agreements Between Aboriginal Communities and Mining Companies: Their use in Canada.* Ottawa: Canadian Environmental Law Association.

Southcott, C. (2009). Introduction: The Social economy and economic development in Northern Canada. *Northern Review*, 30, 3–11.

Southcott, C. ed. (2015). *Northern Communities Working Together: The Social Economy of Canada's North.* Toronto: University of Toronto Press.

Stammler, F. (2010). Our movement to retire the term human dimension from the Arctic science vocabulary. *Northern Notes*, 34, 7–13.

Stammler, F. and Peskov, V. (2008). Building a "culture of dialogue" among stakeholders in North-West Russian oil extraction. *Europe–Asia Studies*, 60(5), 831–849.

Stammler, F. and Wilson, E. (2006). Dialogue for development: An exploration of relations between oil and gas companies, communities, and the state. *Sibirica*, 5(2), 1–43.

Steinberg, P., Tasch, J. and Gerhardt, H. (2015). *Contesting the Arctic: Rethinking Politics in the Circumpolar North*. London: I B Tauris.

Stephenson, S. R., Smith, L. C., Brigham, L. W. and Agnew, J. A. (2013). Projected 21st-century changes to Arctic marine access. *Climatic Change*, 118(3–4), 885–899.

Stewart, E. J., Draper, D. and Johnston, M. E. (2005). A review of tourism research in the polar regions. *Arctic*, 58, 383–394.

Stokes, D.E. (1997). *Pasteur's Quadrant: Basic Science and Technological Innovation*. Washington DC: Brookings Institution.

Stuhl, A. (2016). *Unfreezing the Arctic: Science, Colonialism, and the Transformation of Inuit Lands*. Chicago and London. The University of Chicago Press.

Sulyandziga, P. and Vlassova, T. (2001). Impacts of climate change on the sustainable development of traditional lifestyles of the indigenous peoples of the Russian North: Towards the development of an integrated scheme of assessment. *The Northern Review*, 24, 200.

Sunnari, V. (2010). *"I cannot speak about it": Physical Sexual Harassment as Experienced by Children at School in Northern Finland and Northwest Russia*. Saarbrücken, Germany: VDM Publishing.

Suopajärvi, L., Poelzer, G. A., Ejdemo, T., Klyuchnikova, E., Korchak, E. and Nygaard, V. (2016). Social sustainability in Northern mining communities: A study of the European North and Northwest Russia. *Resources Policy*, 47, 61–68.

Svensson, T. G. (2005). Interlegality, a process for strengthening indigenous peoples' autonomy: The case of the Sámi in Norway. *The Journal of Legal Pluralism and Unofficial Law*, 37(51), 51–77.

Tamnes, R. and Offerdal, K. eds. (2014). *Geopolitics and Security in the Arctic: Regional Dynamics in a Global World*. Abingdon, UK: Routledge.

Tanner, A. (1999). Culture, Social Change, and Cree Opposition to the James Bay Hydroelectric Development. In J.F. Hornig. ed. *Social and Environmental Impacts of the James Bay Hydroelectric Project.*, ed. Montreal QC: McGill-Queen's University Press. pp. 121–140.

Tanner, T., Lewis, D., Wrathall, D., Bronen, R., Cradock-Henry, N., Huq, S. ... Alaniz, R. (2015). Livelihood resilience in the face of climate change. *Nature Climate Change*, 5(1), 23–26.

Taylor, A., Carson, D. B., Ensign, P. C., Huskey, L., Rasmussen, R. O., and Saxinger, G. (2016). *Settlements at the Edge*. Cheltenham: Edward Elgar Publishing.

Tedsen, Elizabeth, Cavalieri, Sandra and Kraemer, R. Andreas. (2014). *Arctic Marine Governance – Opportunities for Transatlantic Cooperation*. Berlin: Springer.

Tennberg, M. (1998). *Arctic Environmental Cooperation: A Study in Governmentality.* Farnham, UK: Ashgate.

Tester, F. J. and Irniq, P. (2008). Inuit Qaujimajatuqangit: Social history, politics and the practice of resistance. *Arctic,* 61(1), 48–61.

Thoreau, H.D. (1854). *Walden; or, Life in the Woods.* Boston MA: Ticknor and Fields.

Tishkov, A. (2012). Тишков, А. А. (2012). Актуальная биогеография как методоло гическая основа сохранения биоразнообразия. Вопросы географии, (134), 15–57.

Tishkov, V. A., Novikova, N. I. and Pivneva, E. A. (2015). Indigenous peoples of the Russian Arctic. *Herald of the Russian Academy of Sciences,* 85(3), 278–286.

Tondu, J.M.E., Balasubramaniam, A.M., Chavarie, L., Gantner, N., Knopp, J.A., Provencher, J.F. …Simmons, D. (2014). Working with Northern communities to build collaborative research partnerships: Perspectives from early career researchers. *Arctic,* 67(3): 419–429.

Trencher, G. P., Yarime, M. and Kharrazi, A. (2013). Co-creating sustainability: Cross-sector university collaborations for driving sustainable urban transformations. *Journal of Cleaner Production, 50,* 40–55.

Trovato, F. and Romaniuk, A. (2014). *Aboriginal Populations: Social, Demographic, and Epidemiological Perspectives.* Edmonton AB: University of Alberta.

Turner, B. L., Kasperson, R. E., Matson, P. A., McCarthy, J. J., Corell, R. W., Christensen, L. … Schiller, A. (2003). A framework for vulnerability analysis in sustainability science. *Proceedings of the National Academy of Sciences,* 100(14), 8074–8079.

Tysiachniouk, M. ed. (2010). Internationalization, trust and multi-stakeholder governance of natural resources. Special Issue, *Journal of Sociology and Social Anthropology.*

Ulturgasheva, O. (2012). *Narrating the Future in Siberia: Childhood, Adolescence and Autobiography Among the Eveny.* New York and Oxford: Berghahn Books.

Ulturgasheva, O., Wexler, L., Kral, M., Allen, J., Mohatt, G. V., Nystad, K. and CIPA Team. (2011). Navigating international, interdisciplinary and indigenous collaborative inquiry: Phase 1 in the circumpolar indigenous pathways to adulthood project. *Journal of Community Engagement and Scholarship,* 4(1), 50.

United Nations. (2002). *Report of the World Summit on Sustainable Development,* Johannesburg, South Africa, A/CONF.199/20, 170.

Usher, P. J. (2003). Environment, race and nation reconsidered: reflections on Aboriginal land claims in Canada. *The Canadian Geographer/Le Géographe canadien,* 47(4), 365–382.

Vaughan, Richard. (2007). *The Arctic - A History.* Chalford, UK: Sutton Publishing.

Vernadsky, W. I. (1945). The biosphere and the noösphere. *American Scientist,* xxii–12.

Vihma, T. (2014). Effects of Arctic sea ice decline on weather and climate: A review. *Surveys in Geophysics,* 35(5), 1175–1214.

Vinogradova, S.N. (2005). *Saami of the Kola Peninsula: Main Trends in the Contemporary Life. Development of the Foundations of the Modern Nature-Use Strategy in the Euro-Arctic Region.* KNC RAS, 424–434. [in Russian]

Vlasova, T. and Volkov, S. (2013). Methodology of socially-oriented observations and the possibilities of their implementation in the Arctic resilience assessment. *Polar Record*, 49(03), 248–253.

Vlassova, T. K. (2006). Arctic residents' observations and human impact assessments in understanding environmental changes in boreal forests: Russian experience and circumpolar perspectives. *Mitigation and Adaptation Strategies for Global Change*, 11(4), 897–909.

Votrin, V. (2006). Measuring Sustainability in the Russian Arctic: An Interdisciplinary Study. PhD thesis. Brussels: Free University of Brussels.

Vulturius, G. and Keskitalo, E. C. H. (2013). Adaptive capacity building in Saxony: Responses in planning and policy to the 2002 flood? In E. C. H. Keskitalo. ed. *Climate Change and Flood Risk Management: Adaptation and Extreme Events at the Local Level*. Cheltenham, UK and Northampton, MA: Edward Elgar Press. pp. 35–66.

Walker, B. and Salt, D. (2006). *Resilience Thinking: Sustaining Ecosystems and People in a Changing World*. Washington DC: Island Press.

Walker, D. A., Webber, P. J., Binnian, E. F., Everett, K. R., Lederer, N. D., Nordstrand, E. A. and Walker, M. D. (1987). Cumulative impacts of oil fields on Northern Alaskan landscapes. *Science*, 238 (4828), 757–761.

Walker, B., Holling, C. S., Carpenter, S. and Kinzig, A. (2004). Resilience, adaptability and transformability in social–ecological systems. *Ecology and Society*, 9(2).

Warde, P. (2011). The invention of sustainability. *Modern Intellectual History*, 8(1), 153–170.

Wassmann, P. and Lenton, T. M. (2012). Arctic Tipping Points in an Earth System Perspective. *Ambio*, 41(1), 1–9. doi:10.1007/s13280-011-0230-9.

Watson, A. and Huntington, O. (2014). Transgressions of the man on the moon: climate change, indigenous expertise, and the posthumanist ethics of place and space, *Geojournal*, 79, 721–736.

Watson Hamilton, J. (2013). Acknowledging and accomodating legal pluralism: An application to the Draft Nordic Saami Convention. In N. Bankes and T. Koivurova. (2013). *The Proposed Nordic Saami Convention: National and International Dimensions of Indigenous Property Rights*. Oxford: Hart Publishing. pp. 45–78.

Wenzel, G. W. (2004). From TEK to IQ: Inuit Qaujimajatuqangit and Inuit cultural ecology. *Arctic Anthropology*, 41(2), 238–250

Wenzel, G. W. (2009). Canadian Inuit subsistence and ecological instability—if the climate changes, must the Inuit? *Polar Research*, 28(1), 89–99.

Wesche, S. and Armitage, D. R. (2010). From the inside out: A multi-scale analysis of adaptive capacity in a Northern community and the governance implications. In Derek Armitage and Ryan Plummer. eds. *Adaptive Capacity and Environmental Governance*. Heidelberg: Springer. pp. 107–32.

Westley, F., Olsson, P., Folke, C., Homer-Dixon, T., Vredenburg, H., Loorbach, D. … van der Leeuw, S. (2011). Tipping toward sustainability: emerging pathways of transformation. *Ambio*, 40(7), 762–780.

White, D. M., Gerlach, S. C., Loring, P., Tidwell, A. C. and Chambers, M. C. (2007). Food and water security in a changing arctic climate. *Environmental Research Letters*, 2(4), 045018.

Wilimovsky, N. J. and Wolfe, J. N. (1966). Environment of Cape Thompson Region, Alaska. Oak Ridge, TN: USAEC Division of Technical Information Extension.

Wilson, E. (2003). Freedom and loss in a human landscape: multinational oil exploitation and the survival of reindeer herding in north-eastern Sakhalin, the Russian Far East. *Sibirica: the Journal of Russia in Asia and the North Pacific*, 3(1), 21–48.

Wilson, E. and Swiderska, K. (2009). *Extractive Industries and Indigenous Peoples in Russia: Regulation, Participation and the Role of Anthropologists*. London: International Institute for Environment and Development.

Wilson, E. and Swiderska, K. (2008). Уилсон, Э. and Свидерска, К. (2008). Горнодобывающая промышленность и коренные народы в России: регулирование, участие и роль антропологов.Этнографическое обозрение, (3).

Wilson, G. N and Alcantara, C. (2012). Mixing politics and business in the Canadian Arctic: Inuit corporate governance in Nunavik and the Inuvialuit settlement region. *Canadian Journal of Political Science*, 45(04), 781–804.

Wilson, S., Pearson, L., Kashima, Y., Lusher, D. and Pearson, C. (2013). Separating adaptive maintenance (resilience) and transformative capacity of social–ecological systems. *Ecology and Society*, 18(1).

Winther, G., Duhaime, G., Kruse, J., Southcott, C. and Aage, H. eds. (2010). *The Political Economy of Northern Regional Development* (Vol. 1). Copenhagen: Nordic Council of Ministers.

Wise, R. M., Fazey, I., Smith, M. S., Park, S. E., Eakin, H. C., Van Garderen, E. A. and Campbell, B. (2014). Reconceptualising adaptation to climate change as part of pathways of change and response. *Global Environmental Change*, 28, 325–336.

Wolf, J., Allice, I., and Bell, T. (2013). Values, climate change, and implications for adaptation: Evidence from two communities in Labrador, Canada. *Global Environmental Change*, 23(2), 548–562.

World Commission on Environment and Development. (1987). *Our Common Future*. Oxford: Oxford University Press.

Worster, D. (1985). *Nature's Economy: A History of Ecological Ideas*. Cambridge: Cambridge University Press.

Wråkberg, U. and Granqvist, K. (2014). Decolonizing technoscience in Northern Scandinavia: the role of scholarship in Sámi emancipation and the indigenization of Western science, *Journal of Historical Geography*, 44, 81–92.

Yagiya, V. S., Kharlampieva, N. K. and Lagutina, M. L. (2015). The Arctic – A new region for China's foreign policy. *International Relations*, (1), 43–52.

Young, Oran R. (1985). The age of the Arctic. *Foreign Policy*, 61, 160–79.

Young, Oran R. (1992a). Sustainable development in the Arctic: The international dimensions. In Oran R. Young. *Arctic Politics: Conflict and Cooperation in the Circumpolar North*. Hanover: University Press of New England. pp. 214–28.

Young, Oran R. (1992b). The Artic in world affairs. In Oran R. Young. *Arctic Politics: Conflict and Cooperation in the Circumpolar North*. Hanover: University Press of New England. pp. 179–89.

Young, Oran R. (1998). *Creating Regimes: Arctic Accords and International Governance*. Ithaca, NY: Cornell University Press.

Young, Oran R. (2009). Whither the Arctic? Conflict or cooperation in the circumpolar North. *Polar Record*, 45(1), 73–82.

Young, Oran R. (2010). Arctic governance - pathways to the future. *Arctic Review on Law and Politics*, (1)2, 164–185.

Young, Oran R. (2011). If an Arctic Ocean treaty is not the solution, what is the alternative? *Polar Record*, 47(04), 327–34.

Young, O. R. (2012). Arctic tipping points: Governance in turbulent times. *Ambio*, 41(1), 75–84.

Young, O. R. (2016). The Arctic Council at twenty: How to remain effective in a rapidly changing environment. UC Irvine Law Review, 6, 99.

Young, T.K. and Bjerregaard, P. (2008). *Health Transitions in Arctic Populations*. Toronto: University of Toronto.

Young, O. R. and Kankaanpa, P. (2012). The effectiveness of the Arctic Council. *Polar Research*, 31, 1–14.

Young, O., Kullerud, L., Southcott, C., Snellman, O., Cochran,P., Fjellheim, R. ... Trondheim, G. (2005). *ICARP II Science Plan 11: Arctic Science in the Public Interest*. Copenhagen, Denmark..http://www.aosb.org/icarp_ii/index.html).

Young, T.K., Rawat, R., Dallmann, W., Chatwood, S. and Bjerregaard, P. (2012). *Circumpolar Health Atlas*. Toronto: University of Toronto.

Zamyatina N. and Pelyasov, A. (2015). *Innovation Search in Monoprofile Towns: Blockages of Development, New Industrial Policy and Plan of Action*. Moscow: Lanand.

Zellen, B. (2013). *The Fast-Changing Arctic*. Edmonton AB: University of Calgary Press.

Ziker, J. P. (2012). Resilience of domestic groups and communities on the Lower Enisei River throughout the twentieth century. *Sibirica*, 11(1), 1–42.

Index